Water Science and Technology Library

Volume 107

The aim of the *Water Science and Technology Library* is to provide a forum for dissemination of the state-of-the-art of topics of current interest in the area of water science and technology. This is accomplished through publication of reference books and monographs, authored or edited. Occasionally also proceedings volumes are accepted for publication in the series. *Water Science and Technology Library* encompasses a wide range of topics dealing with science as well as socio-economic aspects of water, environment, and ecology. Both the water quantity and quality issues are relevant and are embraced by *Water Science and Technology Library*. The emphasis may be on either the scientific content, or techniques of solution, or both. There is increasing emphasis these days on processes and *Water Science and Technology Library* is committed to promoting this emphasis by publishing books emphasizing scientific discussions of physical, chemical, and/or biological aspects of water resources. Likewise, current or emerging solution techniques receive high priority. Interdisciplinary coverage is encouraged. Case studies contributing to our knowledge of water science and technology are also embraced by the series. Innovative ideas and novel techniques are of particular interest.

Comments or suggestions for future volumes are welcomed.

Vijay P. Singh, Department of Biological and Agricultural Engineering & Zachry Department of Civil and Environment Engineering, Texas A &M University, USA
Email: vsingh@tamu.edu

All contributions to an edited volume should undergo standard peer review to ensure high scientific quality, while monographs should also be reviewed by at least two experts in the field.

Manuscripts that have undergone successful review should then be prepared according to the Publisher's guidelines manuscripts: https://www.springer.com/gp/authors-editors/book-authors-editors/book-manuscript-guidelines

More information about this series at http://www.springer.com/series/6689

Ramakar Jha · Vijay P. Singh · Vivekanand Singh ·
L. B. Roy · Roshni Thendiyath
Editors

Water Resources Management and Reservoir Operation

Hydraulics, Water Resources and Coastal Engineering

Springer

Editors
Ramakar Jha
Department of Civil Engineering
National Institute of Technology Patna
Patna, India

Vivekanand Singh
Department of Civil Engineering
National Institute of Technology Patna
Patna, India

Roshni Thendiyath
Department of Civil Engineering
National Institute of Technology Patna
Patna, India

Vijay P. Singh
Department of Biological and Agricultural Engineering
Texas A&M University
College Station, TX, USA

L. B. Roy
Department of Civil Engineering
National Institute of Technology Patna
Patna, India

ISSN 0921-092X ISSN 1872-4663 (electronic)
Water Science and Technology Library
ISBN 978-3-030-79399-9 ISBN 978-3-030-79400-2 (eBook)
https://doi.org/10.1007/978-3-030-79400-2

This Springer imprint is published by the registered company Springer Nature Switzerland AG
The registered company address is: Gewerbestrasse 11, 6330 Cham, Switzerland

Contents

1 **Integrated Water Resources Management of Thatipudi Command Area, Vizianagaram, Andhra Pradesh** 1
K. Kalyani and K. V. Jayakumar

2 **Hydrological Modelling to Study the Impacts of Climate and LULC Change at Basin Scale: A Review** 13
Dinu Maria Jose, Waleed Makhdumi, and Gowdagere Siddaramaiah Dwarakish

3 **Water Resource Management for Coal-Based Thermal Power Plant** 27
S. A. Nihalani, Y. D. Mishra, and A. R. Meeruty

4 **Evaluation of Reservoir Sedimentation Using Satellite Data—A Case Study** 41
Beeram Satya Narayana Reddy and S. K. Pramada

5 **Regionalisation of Watersheds Using Fuzzy C Means Clustering Algorithm in the West Flowing River of Kerala** 51
Thottungal Krishnankutty Drissia, Vinayakam Jothiprakash, and Alayil Bahuleyan Anitha

6 **Analysis of Relationship Between Landslides and Rainfall in Karwar, Uttara Kannada District, Karnataka, India** 65
G. Thejashree, K. N. Lokesh, and G. S. Dwarakish

7 **Optimal Cropping Pattern of Kulsi River Basin, Assam, India Using Simulation and Linear Programming Model** 77
Jyotismita Taye, Bibhash Sarma, and Abhijit D. Lade

8 **Comparison of Flux Footprint Models to a Mixed Fetch Heterogeneous Cropland System** 91
Shweta Kumari and K. B. V. N. Phanindra

9 Water Resources Assessment Issues and Application of Isotope Hydrology in North East India 103
Prem Ranjan, Pankaj Kumar Pandey, Vanita Pandey, and Pema Tshering Lepcha

10 Water Hammer Analysis for Pipe Line Network Using HAMMER V8i 117
Ajmal Hussain, Muhammad Mustafa, S. M. Ahbar Warsi, and Sumit Kumar

11 Dam Break Flood Routing and Inundation Mapping Using HEC-RAS and HEC-GeoRAS 129
A. Bharath, Anand V. Shivapur, and C. G. Hiremath

12 Suitability and Performance of Present Irrigation System in Kokernag, Jammu and Kashmir 139
Dar Himayoun, Roshni Thendiyath, and Jahangeer Saleem

13 Linking of Sediment Yield Pattern with Rainfall and Land-Use Land-Cover Changes Within Burhanpur Sub-catchment, India 155
S. R. Resmi, P. L. Patel, and P. V. Timbadiya

14 Assessment of Probable Maximum Flood (PMF) Using Hydrologic Model for Probable Maximum Precipitation in Maithon Watershed 165
Bhanu Sharma and Kalyan Kumar Bhar

15 Simulating Failure of Indravati Dam Using Mike 11 and the Propagation of Breached Outflow 177
Aditya Harikumar, Sachin Dhiman, and K. C. Patra

16 Optimization of Water Allocation for Ukai Reservoir Using Elitist TLBO 191
Vijendra Kumar and S. M. Yadav

17 Prediction of Reservoir Submerged Sediment Density 205
Y. C. Jabbar and S. M. Yadav

18 Micro-hydro Power Generation in India—A Review 219
Aparna M. Deulkar, Vivek S. Chavhan, and Pankaj R. Modak

19 Runoff Simulation and Irrigation Water Requirement for Barman Command 227
A. Vishwakarma, M. K. Choudhary, and M. S. Chauhan

20 Nonlinear Regression Analysis Between Discharge and Head for Piano Key Weirs with Increasing Developed Length (L/W) Ratio and Constant Channel Width 241
Amiya Abhash and K. K. Pandey

21 Grey Water Characterization and Its Management 251
Sarosh Alam Ghausi and Mohd Muzzammil

22 Intelligent Operation of Hirakud Reservoir Using Metaheuristic Techniques (PSO and TLBO) 263
Pooja Patnaik and Prakash Ch. Swain

23 Agricultural Water Management and Groundwater Recharging Using Vadose Zone Modelling 277
Anooja Thomas, Vivekanand Singh, and Brijesh Kumar Yadav

About the Editors

Ramakar Jha is Chair Professor in the Department of Civil Engineering and has 30 years of experience in the field of hydrology and water resources engineering. He is presently working as Chair Professor in the Department of Civil Engineering, National Institute of Technology (NIT) Patna, India, which is a premier institute in India under the Ministry of Human Resource Development, Government of India. He served at various levels from Scientist-B to Scientist-E1 at National Institute of Hydrology (NIH), Roorkee, India, and as Professor in the Department of Civil Engineering, NIT Rourkela. He has worked and working as Country Coordinator of UNESCO-GWADI and Principal Investigator for many international (EU-FP7, DAAD, ADB, AUS-Aid) and national research and consultancy projects (ISRO, DST, MoWR, MHRD). Moreover, he served as Chair for many administrative positions and received a couple of international and national awards for research papers. Presently, he is working as Dr. Rajendra Prasad Chair for Water Resources under the Ministry of Water Resources, Government of India, in the Department of Civil Engineering, NIT Patna, Bihar.

Vijay P. Singh Texas A&M Professor of Indian origin is receiving a prestigious award for his world-renowned work on water. He is receiving the 2013 Lifetime Achievement Award from the American Society of Civil Engineers-Environmental and Water Resources Institute, otherwise known as the ASCE-EWRI. The award is in recognition of his work in the field of hydrology, which is the study of water in all aspects, such as quality, distribution, preservation, and transportation. Some of the work he has done has even created an entire new branch of hydrology—called entropic hydrology—that is connected to the study of entropy, which means essentially the study of order and disorder as it relates to the physical universe. His work is considered fundamental for flood planning and water modeling around the world. Since earning his doctorate degree, he has held teaching positions in some of the most well-known universities in the USA. He was Associate Research Professor of civil engineering at George Washington University from 1977 to 1978, Associate Professor of civil engineering at Mississippi University from 1978 to 1981, and Adjunct Professor as well as Coordinator of the Environmental and Water Resources Systems Engineering Program at Louisiana State University from 1999 to 2006 and

2001 to 2006, respectively. He joined Texas A&M University in 2006, where he currently wears a number of different hats. He is Professor of biological and agricultural engineering, Professor of civil and environment engineering, and Caroline and William N. Lehrer Distinguished Chair in Water Engineering (Hydrology). He has authored or edited around 10 published works in the fields of engineering and hydrology.

Vivekanand Singh is Professor in the Department of Civil Engineering and has 28 years of experience in the field of river hydraulics, groundwater, and water resources engineering. He has published research papers in international journals including ASCE Journal. He has done national projects and consultancy during his work at NIT Patna. Prior to this, he was Scientist at National Institute of Hydrology, Roorkee. He organized several summer courses, conferences, and workshops. He has been working as Editor in some Indian journal.

L. B. Roy is Professor in the Department of Civil Engineering and has 33 years of experience in the field of water resources engineering as well as geotechnical engineering. He has been serving the Water Resources Engineering Department for long time and carried out various field-based research activities for different river systems of the region with special emphasis to floods. He has been involved in many research projects and published several research papers in peer-reviewed journals. He has been working for Intellectual Property Rights (IPR) and received awards for various activities.

Roshni Thendiyath is Assistant Professor in the Department of Civil Engineering and has 9 years of experience in the field of water resources engineering. She earned her Ph.D. in civil engineering at the University of Pisa, Italy. After obtaining Ph.D., she had joined National Institute of Technology Calicut, India, as Adhoc Faculty and later joined National Institute of Technology Patna as Assistant Professor. She has supervised two Ph.D. and more than 30 master's theses. Furthermore, she has been awarded DST-SERB project for the topic Two-Phase Flows and Water Quality in Rivers, funded by the Department of Science and Technology, MHRD, India. Results obtained from her research have been published nearly 30 papers in international journals and more than 15 papers in international conferences and three chapters. She is active in a variety of professional bodies, and she has organized numerous workshops and conferences in her academic career.

Chapter 1
Integrated Water Resources Management of Thatipudi Command Area, Vizianagaram, Andhra Pradesh

K. Kalyani and K. V. Jayakumar

Abstract IWRM provides ideas to help us make social choices about water allocation, the sustainability of water resources and the infrastructure use to manage the resources establishment of an overall water policy and laws which use the basin as the scale of management. It is widely recognised that better water management and governance is the key to developing a sustainable, efficient and equitable water sector. The present study area is Thatipudi reservoir and its command area. The command area is characterised by highly fluctuating rainfall. This paper discusses on preparing GIS maps, water requirement for command area using crop water requirements and irrigation scheduling software, storage capacity of the reservoir and cropping pattern to implement IWRM on command area. Highly fluctuating rainfall and corresponding run-off water was analysed considering the irrigation requirements, conveyance efficiency, domestic water requirements and water release into the sea during the floods. The present study reveals that the current capacity of the reservoir is insufficient to cater the allocation needs and thus necessitates the increase in storage capacity to cater the water needs in dry season.

Keywords Integrated Water Resource Management (IWRM) · Irrigation scheduling · Crop water requirements and reservoir capacity

1.1 Introduction

IWRM is a logical and appealing concept, its basis is that many different uses of water resource are interdependent. Safe water supply and environmental sanitation are essential for protecting the environment, improving health and reduction of poverty. Despite increases in awareness of Integrated Water Resources Management (IWRM)

K. Kalyani (✉)
Department of Civil Engineering, Andhra University, Visakhapatnam, AP 530003, India

K. V. Jayakumar
Civil Engineering, National Institute of Technology, Warangal, Telangana 506004, India
e-mail: kvj@nitw.ac.in

R. Jha et al. (eds.), *Water Resources Management and Reservoir Operation*,
Water Science and Technology Library 107,
https://doi.org/10.1007/978-3-030-79400-2_1

and a number of studies that focus on the concept, still there exist a few in depth scientific studies on the rationale behind the principles and concepts. Management is used in its broadest sense. It emphasizes that one must not only focus on development of water resources but that must manage water development in a way that ensures long term sustainable use for future generations. IWRM is a process without formulas and at the same time highlighted in guidelines produced by the Global Water Partnership. IWRM attempts to put the Dublin Principles into practice, emphasizing the ideas of integration, participation, and economic and financial sustainability and with the basin as the unit for decision making.

Several large international conferences and agreements had particular influence over its formation on Stockholm conference (1972) on the environment, Mar del Plata Conference (1977) and the Dublin Conference (1992) held in preparation for the United Nations Conference on Environment and Development (UNCED). Agarwal et al. (2000) provides an extensive overview of established methods and approaches that could be employed in water resources management. They comprise a wide range of instruments used in assessing water resources, regulation, economic management, conflict resolution, communication, and new technology.

IWRM provides ideas to help us consider how we can best make social choices about water allocation and access as well as the sustainability of water resources and the infrastructure use to manage those resources indicates by Giordano and Shah (2014). Van Der Keuret al. (2010) identifies the characteristics and uncertainty as it occurs in the different stages of the IWRM process with respect to sources, nature and type of uncertainty and a common terminology that honour the most important aspects from natural and social sciences and its application to the entire IWRM process. Warwick and Haness (1994) observed that the conventional hydrological data are inadequate for the purpose of design and operation of water resources system. In such cases GIS are of great use for estimation of relevant hydrologic data. GIS can be used for determination of catchment characteristics.

Integration across policy sectors is a core issue of IWRM. Grigg (2008) agree that IWRM should coordinate the development and management of water, land and other resources in order to improve economic and social welfare, social equity, and environmental sustainability (GWP 2004). This recognizes the inherent interdependencies in nature and in the economic and social sectors. The GWP (2004) advocates that the IWRM approach seeks to address a country's key water-related development problems, water for health, for food, for energy, for environment more effectively and efficiently than is possible using traditional approaches.

Rees et al. (2008) argued that water governance reform need to actively recognize and incorporate the financial dimension. Likewise, efforts to attract the additional finance so badly needed by the sector have to be accompanied by governance reforms to ensure that the funds are employed efficiently and that the financial flows can be sustained.

1.2 Methodology

The study area details and the methodology adopted for the study area (Fig. 1.1) were gathered and used in planning of water management programs.

1.2.1 Study Area

Thatapudi reservoir is constructed across the Gosthani River in Vizianagaram district of Andhra Pradesh. The district is a part of the northern coastal plains of Andhra Pradesh and it is situated within the geographical coordinates of 17°–15′ and 19°–15′ of the Northern latitude and 83°–0′ to 83°–45′ of the Eastern longitude. The reservoir was built during the period 1963–68 to benefit an ayacut (culturable command area) of 4868 hectares spread over three Mandals viz., Gantyada, Jami and S. Kota and provide water for irrigation purpose and water supply to Visakhapatnam Municipal Corporations at the rate of 22.70 MLD.

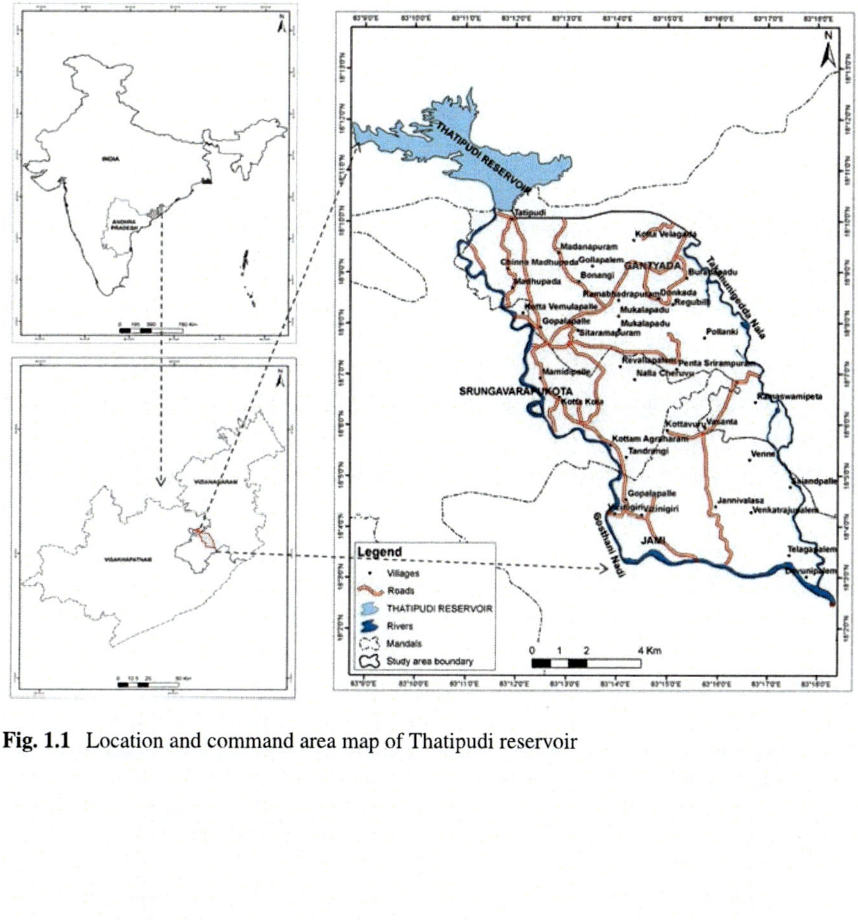

Fig. 1.1 Location and command area map of Thatipudi reservoir

The Gosthani's river water is diverted for agricultural and industrial purposes and the river is the chief source of drinking water to the cities of Vizianagaram and Visakhapatnam. Several infiltration wells have been sunk on the Gosthani river bed to extract water especially during the summer months. Water that released from canals is useful for villages and tribal population. It is also a picturesque tourist location.

1.2.2 Data Collection

Data for the study were available from various sources like (i) Thatipudi reservoir (ii) Indian Meteorological Department and its Cyclone section (iii) Municipal Corporation of Visakhapatnam, and (iv) Irrigation and Command Area Development Department (I&CADD).

The required data (1997 onwards) were collected from various departments in Vizianagaram and Visakhapatnam districts. This study has objectives like estimation of storage capacity of the reservoir, water requirement for command area using crop water requirements and irrigation scheduling software, various maps using GIS and cropping pattern to implement IWRM on command area.

1.3 Analysis

The maps acquired from survey of India, are scanned and the final maps are presented in ARCGIS as mandals and village's maps, location and drainage maps. From these maps water management phases are prepared from mandals and village's maps cropping pattern study proceeds to water requirement study than water demand for crop seasonally can be identified. Now from location and drainage maps annual rain fall data and storage capacity of reservoir are calculated. If water in the reservoir is excess after meeting the demand management plan develops. If it is deficit after meeting the demand then water developing and ground water extracting process should be developed. The analysis is presented in flow chart as shown in Fig. 1.2.

1.3.1 Reservoir Storage Capacity

The current capacity of the reservoir is determined by Graphical method, using mass curves. The capacity required for a reservoir depends upon the inflow available and demand. Storage may not be required if the available inflow in the river is greater than the demand. Larger capacity of the reservoir is required for higher demand if the inflow into the reservoir is less.

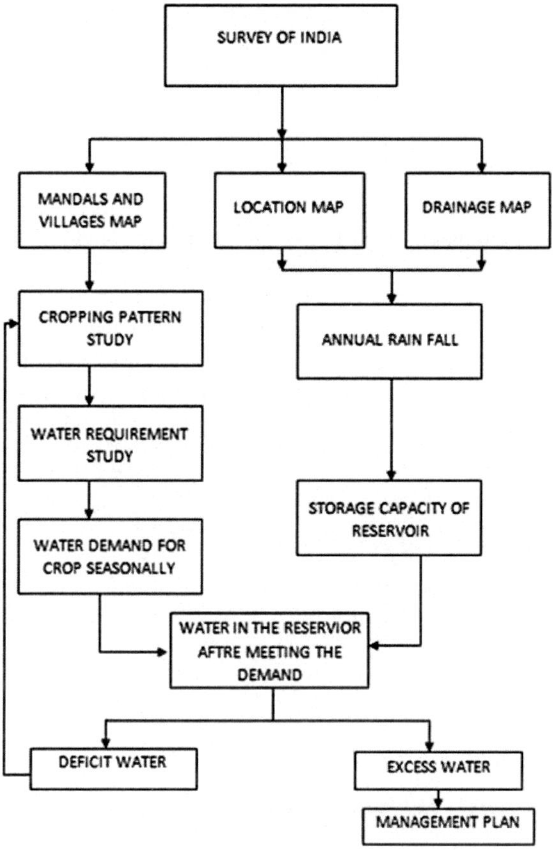

Fig. 1.2 Flow chart of management plan

$$S = \left(\sum I_t - \sum O_t\right) = \sum (I_t - O_t) \tag{1.1}$$

where,

S = Net storage

I_t = Volume of reservoir inflow

O_t = Volume of reservoir demand.

1.3.2 Irrigation Purpose

The village lands are irrigated by Thatipudi reservoir and a total of 155.54 ha (385 acres) are irrigated under it in the village. According to general public, the reservoir water is not sufficient for crops and also due to very poor maintenance of irrigation canals. Major crops grown are paddy, vegetables and pulses.

1.3.3 Water Supply

From Thatipudi reservoir 22.70 MLD of water is supplied to Visakhapatnam. Municipal Corporation of Visakhapatnam apportions the water for industries and for drinking water supply. Gosthani water supply scheme executed by the military in 1942 was taken over by the Municipality. In order to make further improvements the authorities executed Gambhiram Gedda scheme to supply 15 MLD in 1957 to supply water during one season. To meet the increasing water demand the authorities executed a scheme in 1967 to supply 22.70 MLD from the Thatipudi reservoir.

1.3.4 Preparation of GIS Maps

The maps obtained from the Survey of India are scanned and then digitized in Auto CAD and the digitized maps are further geo-referenced and the areas are calculated using ARCINFO and the final maps are presented in ARCGIS, which are the command area map (Fig. 1.1) and drainage map (Fig. 1.3) respectively.

1.3.4.1 Delineation of Thematic Maps

The toposheets are used for determination of command area of all drains of Thatipudi reservoir. In the present study command area is determined using GIS software. The command area of Thatipudi reservoir is 9.61243 km^2 and study area boundary is 107.398 km^2. The catchment area of Gosthani river areas is 3.47014 km^2 and another river found in these toposheets maps i.e. TakanunigeddaNala, a tributary of Gosthani River has a catchment area of 1.19331 km^2. 37 villages are found under Thatipudi command area.

1.4 Results

Annual graphical rainfall analysis was carried for Thatipudi command area with the data from past year and presented.

The storage capacity is calculated by graphical method using mass curve and graphically represented.

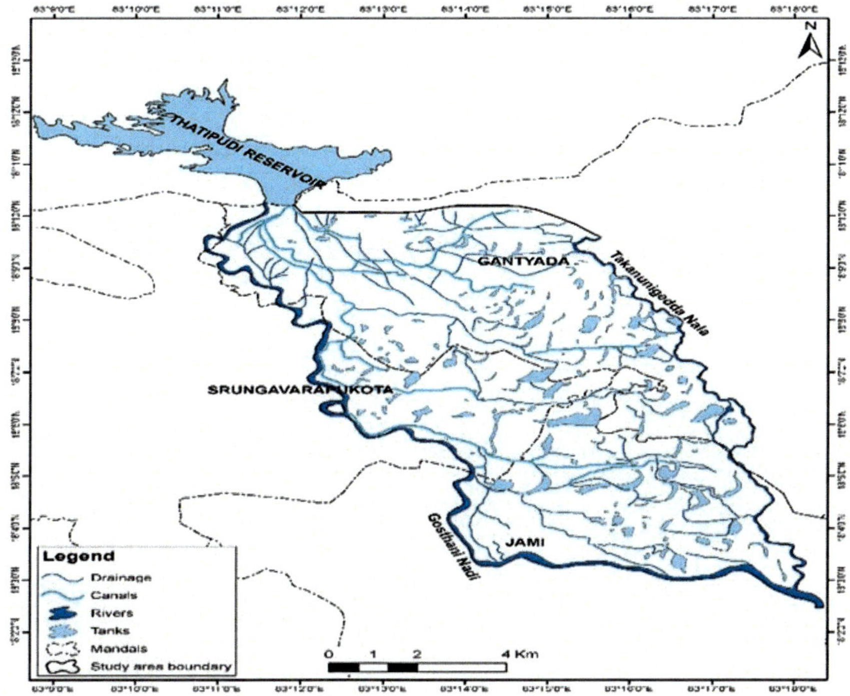

Fig. 1.3 Drainage network map of Thatipudi reservoir

1.4.1 Irrigation in Kharif Season

1.4.1.1 Water Requirement for Paddy Crop

Thatipudi Reservoir scheme which benefits an Ayacut of irrigated area spread over three Mandals viz., Gantyada, Jami and S. Kota and major irrigation projects in Vizianagaram District. Calculation of water requirements for Paddy irrigated area 8875 ha in the study area in Kharif season (Fig. 1.4; Table 1.1).

1.4.1.2 Water Requirement for Pulses Crop

Irrigated dry crop area is 460 ha in this command area in kharif season and most of the crops are pulses and ground nuts. Water requirement is calculated for pulses over an irrigated area of 300 ha (Table 1.2).

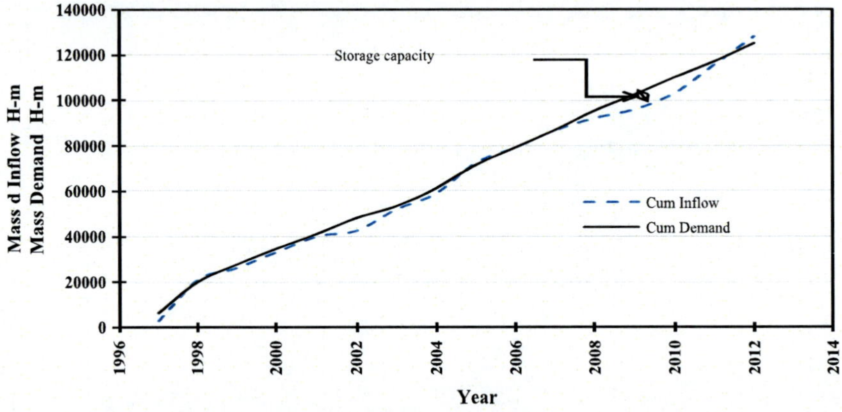

Fig. 1.4 Mass curve of the Thatipudi reservoir

1.4.1.3 Water Requirement for Ground Nut Crop

Water requirement is calculated for ground nuts over an irrigated area of 160 ha in kharif season in Thatipudi command area (Table 1.3).

1.4.2 Irrigation in Rabi Season

1.4.2.1 Water Requirement for Maize Crop

Calculation of water requirements in rabi season for irrigated dry area 4498 ha, in that maize irrigated area is 2500 in this study area (Table 1.4).

1.4.2.2 Water Requirement for Vegetables

Calculation of water requirements in rabi season for irrigated dry area 4498 ha, in that vegetables irrigated area is 1998 in this study area (Table 1.5)

1.5 Conclusions

Given the variation in the nature and complexity of the issues, the water and water-related problems have to be approached at different levels. The study was undertaken with a view to implement IWRM for Thatipudi reservoir command area and the following conclusions are presented.

Table 1.1 Water requirement for paddy crop

Month	Decade[a]	Stage[b]	Crop coefficient (K_c)	E_{tc} (mm/day)	E_{tc} (mm/decade)	Effective rainfall (mm/dec)	Irrigation required (mm/dec)
June	1	Initial	1.10	5.34	53.40	1.30	5.30
	2	Initial	1.10	5.31	53.10	5.00	48.10
	3	Development	1.10	5.27	52.70	18.70	34.00
July	1	Development	1.09	5.20	52.00	39.70	12.30
	2	Development	1.08	5.10	51.00	53.40	0.00
	3	Middle	1.07	5.03	50.30	45.50	9.90
Aug.	1	Middle	1.06	5.01	50.10	33.80	16.30
	2	Middle	1.06	5.00	50.00	27.60	22.40
	3	Middle	1.06	4.87	48.70	28.10	25.50
Sept.	1	Middle	1.06	4.74	47.40	30.50	16.90
	2	Middle	1.06	4.61	46.10	30.90	15.20
	3	Later	1.06	4.38	43.80	25.60	18.20
Oct.	1	Later	1.04	4.07	40.70	19.90	20.90
	2	Later	1.01	3.75	37.50	15.30	22.20
	3	Later	0.98	3.45	34.50	6.50	13.90
					662.80	381.70	2810

Water requirement for paddy area = 24,938,750 m^3

Table 1.2 Water requirement for Pulsus

Month	Decade[a]	Stage[b]	Crop coefficient (K_c)	E_{tc} (mm/day)	E_{tc} (mm/dec)	Effective rainfall (mm/dec)	Irrigation Required (mm/dec)
June	1	Initial	0.40	1.94	19.40	1.30	1.90
	2	Initial	0.40	1.93	19.30	5.00	14.30
	3	Development	0.40	1.93	19.30	18.70	0.60
July	1	Development	0.54	2.57	25.70	39.70	0.00
	2	Development	0.75	3.56	35.60	53.40	0.00
	3	Middle	0.97	4.60	46.00	45.50	5.00
Aug.	1	Middle	1.04	4.92	49.20	33.80	15.40
	2	Middle	1.04	4.92	49.20	27.60	21.50
	3	Middle	1.04	4.79	47.90	28.10	24.60
Sept.	1	Later	1.02	4.56	45.60	30.50	15.10
	2	Later	0.75	3.25	32.50	30.90	1.60
	3	Later	0.45	1.88	18.80	17.90	0.00
					408.50	332.40	100.10

Water requirement for pulses area ℑ 300,300 m^3

Table 1.3 Water requirement for Ground Nut

Month	Decade[a]	Stage[b]	Crop coefficient (K_c)	E_{tc} (mm/day)	E_{tc} (mm/dec)	Effective rainfall (mm/dec)	Irrigation Required (mm/dec)
June	1	Initial	0.40	1.94	19.40	1.30	1.90
	2	Initial	0.40	1.93	19.30	5.00	14.30
	3	Initial	0.40	1.92	19.20	18.70	0.50
July	1	Development	0.44	2.09	20.90	39.70	0.00
	2	Development	0.61	2.89	28.90	53.40	0.00
	3	Development	0.81	3.80	38.00	45.50	0.00
Aug.	1	Middle	0.99	4.69	46.90	33.80	13.10
	2	Middle	1.05	4.93	49.30	27.60	21.60
	3	Middle	1.05	4.80	48.00	28.10	24.70
Sept.	1	Middle	1.05	4.67	46.70	30.50	16.10
	2	Middle	1.05	4.54	45.40	30.90	14.50
	3	Later	0.97	4.00	40.00	25.60	14.40
Oct.	1	Later	0.75	2.95	29.50	19.90	9.70
	2	Later	0.57	2.11	14.80	10.70	0.00
					428.30	370.70	130.70

Water requirement for ground nuts area ℑ 209,120 m^3

Table 1.4 Water requirement for maize crop

Month	Decade[a]	Stage[b]	Crop coefficient (K_c)	E_{tc} (mm /day)	E_{tc} (mm/dec)	Effective rainfall (mm/dec)	Irrigation required (mm/dec)
Dec	2	Initial	0.30	0.81	8.10	7.70	0.00
	3	Initial	0.30	0.82	8.20	8.50	0.40
Jan	1	Development	0.36	0.99	9.90	0.20	9.70
	2	Development	0.57	1.58	15.80	0.00	15.80
	3	Development	0.79	2.30	23.00	0.00	25.20
Feb.	1	Middle	1.00	3.04	30.40	1.40	29.00
	2	Middle	1.05	3.31	33.10	2.00	31.10
	3	Middle	1.05	3.63	36.30	1.40	27.60
Mar.	1	Middle	1.05	3.94	39.40	0.10	39.40
	2	Later	1.04	4.25	42.50	0.00	42.50
	3	Later	0.88	3.74	37.40	0.50	40.70
Apr.	1	Later	0.64	2.81	28.10	14.00	14.10
	2	Later	0.43	1.96	19.60	16.70	0.00
					331.80	52.30	275.50

Water requirement for maize irrigated area ℑ 6,887,500 m^3

Table 1.5 Water requirement for vegetables

Month	Decade[a]	Stage[b]	Crop coefficient (K_c)	E_{tc} (mm/day)	E_{tc} (mm/dec)	Effective rainfall (mm/dec)	Irrigation Required (mm/dec)
Dec.	2	Initial	0.70	1.88	18.80	7.70	4.90
	3	Initial	0.70	1.90	19.00	8.50	12.40
Jan.	1	Development	0.73	1.99	19.90	0.20	19.80
	2	Development	0.81	2.26	22.60	0.00	22.60
	3	Development	0.91	2.63	26.30	0.00	28.90
Feb.	1	Middle	0.97	2.94	29.4	1.4	28
	2	Middle	0.97	3.06	30.60	2.00	28.60
	3	Middle	0.97	3.35	33.50	1.40	25.40
Mar.	1	Later	0.95	3.59	35.90	0.10	35.80
	2	Later	0.89	3.60	36.00	0.00	32.40
					272.00	21.20	238.80

Water requirement for vegetables irrigated area ℑ 4,771,224 m^3

- The annual average rainfall in the study area is 1157.34 mm. The rainfall is highly fluctuating in the study area and it is identified the need for Integrated Water Resources Management. Thatipudi reservoir storage capacity was found to be 97.93 M.Cum.
- It is observed from the above data that most the inflow into the reservoir is less than the demand. So it is indicating that proper integrated water resources management plan can be developed for extracting additional water for various other developmental purposes.
- It can be seen that water requirement for irrigation throughout the year is around 37.11 M.Cum and considering conveyance efficiency also, it would be 53.01 M.Cum. From the data available it can be seen that 51.46 M.Cum of water is released for irrigation purpose by I & CAD department.
- Water Supplied to Visakhapatnam city is found to be 18.00 M.Cum against a requirement of 97.93 M.Cum
- By implementing various water resource techniques, water can be stored and supplied to various users to fulfill the demand.
- By implementing IWRM in Thatipudi command area water resource management can be developed and water utilisation will be developed in various factors in right way.

References

Agarwal A, Delos Angeles MS, Bhatia R, Chéret I, Davila-Poblete S, Falkenmark M, Wright A (2000) Integrated water resources management. Global Water Partner 64(8):109–117

Giordano M, Shah T (2014) From IWRM back to integrated water resources management. Int J Water Resour Dev 36(6):1–13

Grigg NS (2008) Integrated water resources management: balancing views and improving practice. J Water Int 33(3):279–292

Rees JA, Winpenny J, Hall AW (2008) Water financing and governance. Global Water Partnership, No. 12

Van der Keur P, Brugnach M, Dewulf ARPJ, Refsgaard JC, Zorilla P, Poolman M, Mysiak J (2010) Identifying uncertainty guidelines for supporting policy making in water management illustrated for Upper Guadiana and Rhine Basins. J Water Resour Manag 24(14):3901–3938

Warwick JJ, Haness SJ (1994) Efficacy of ARC/INFO, GIS application to hydrological modeling. J WRD and ASCE 120(3):366–379

Chapter 2
Hydrological Modelling to Study the Impacts of Climate and LULC Change at Basin Scale: A Review

Dinu Maria Jose, Waleed Makhdumi, and Gowdagere Siddaramaiah Dwarakish

Abstract Two key factors which alter the hydrological system are human activity and climate change. They affect the spatial and temporal distribution of water and have unequal influence on hydrological system. Hence, the study of influence of climate change and human activity is of immense importance for planning schemes and strategies for water resources management. As a consequence of urbanization, there has been a change in Land Use and Land Cover (LULC), which resulted in an increase of surface runoff and degradation in water quality. Hydrological modelling is an important tool which is used by researchers all over the world for analyzing the influence of climate change and Land Use Land Cover Change (LULCC) on natural water sources and for predicting potential future impacts from various possible scenarios. Based on the future climate change scenarios, numerous basins may probably experience variation in the mean hydrological characteristics as well as in the magnitude and frequency of extreme events. The main goal of this paper is to give a brief review of studies done to evaluate the impacts of climate change and LULCC on hydrological regime. There is a widespread usage of Landsat series of data for hydrological modelling studies based on the period of analysis. And majority of the hydrological modelling studies used semi-distributed models like SWAT for their analysis as these aren't much data intensive but gives reasonably accurate results. Based on the results of various modelling studies, it can be concluded that, depending on the characteristics of watershed and scenario assumptions, the combined effects of climate change and LULCC may ameliorate or deteriorate each other's influence. These effects may vary with the season or land use classes involved in change. Hydrological modelling integrating future climate change and LULCC scenarios can be an effective tool in planning future water resource management strategies.

Keywords Climate change · Hydrological modelling · Urbanization · Land use land cover change · Scenarios

D. M. Jose (✉) · W. Makhdumi · G. S. Dwarakish
Department of Water Resources and Ocean Engineering, National Institute of Technology Karnataka, Surathkal, Mangaluru, India

R. Jha et al. (eds.), *Water Resources Management and Reservoir Operation*, Water Science and Technology Library 107,
https://doi.org/10.1007/978-3-030-79400-2_2

2.1 Introduction

Water is one among the crucial components of the environment and needs proper management to accomplish its sustainable utilization in the future. Climate change can affect the water availability on a large scale. Land use land cover changes like urban growth can also alter the hydrological cycle affecting processes like evapotranspiration, infiltration, and surface and subsurface runoff (Dey and Mishra 2017). These two influencing forces can affect the hydrological components at global, continental, regional and basin scales in all parts of the world (Praskievicz and Chang 2009). Hydrological modelling is an important tool which is used by researchers all over the world for analyzing the influence of climate change and Land Use Land Cover Change (LULCC) on natural water sources and for predicting potential future impacts from various possible scenarios. Initially this paper gives a brief introduction to various hydrological models. Later on, the studies on individual and combined impacts of land use land cover and climate change are described. Finally, model comparison studies and conclusions are presented.

2.2 Models

Hydrological models are widely used across the world for studying the effects of various processes on water resources and thus for predicting potential future impacts on the same (Praskievicz and Chang 2009). The models can be mainly classified based on the physical and spatial structure (Krysanova et al. 2015). The models can be grouped as distributed, semi-distributed and lumped based on the representation of spatial heterogeneity at watershed scale (Dwarakish and Ganasri 2015). Distributed and semi-distributed models capture the spatial variability better than the lumped models (Bormann et al. 2009). Depending on the process description, models can be categorized as conceptual, empirical and physical. Based on the consideration of randomness, models are classified as stochastic and deterministic models. Former involves randomness while latter doesn't (Dwarakish and Ganasri 2015).

2.3 Hydrological Modelling in LULC and Climate Change Impact Studies

Latitude, topography, geology, and land use affect the hydrological impacts of climate in a basin. The rise in the surface runoff and undesirable high and low storm hydrograph are significant impacts of land use land cover change (Praskievicz and Chang 2009). The changes in climate and land use land cover (LULC) are interconnected and are capable of modifying hydrological processes (Legesse et al. 2003; Li et al. 2009). Nevertheless, the influence of climate and LULC change on hydrological regime are frequently researched separately. For example, there are researches done to find out the influence of LULCC on evaporation losses (Cristina et al. 2015), infiltration rates (Weatherhead and Howden 2009), and on runoff volume (Hundecha and Bárdossy 2004). And, there are studies which predict how climate change affects streamflow (Mudbhatkal et al. 2017; Mudbhatkal and Amai 2017; Piao et al. 2010; Treesa et al. 2017), groundwater recharge (Soro et al. 2017), sediment load (Rodríguez-Blanco et al. 2016) and water quality (Glavan et al. 2015).

2.3.1 Impacts of LULCC on Hydrology

There are many studies which evaluate the impact of LULC change on various temporal and spatial scales (Fan and Shibata 2015; Zhou et al. 2013). As a result of anthropogenic activities there has been subsequent increase in impervious area. This leads to the alteration of the water balance of the catchment, with increase in runoff, decrease in evapotranspiration and groundwater recharge (Leopold and Dunne 1978). Factors like altitude, slope, distance from the river, type of agricultural practices, type of soil and magnitude of erosion, frequency of drought and flood, population density and distance from a built-up area affect the rate of LULC change (Lin et al. 2009). The magnitude of impacts depend on the several factors like soil depth (FAO 2008), precipitation events (FAO 2008), spatial layout of deforestation areas (National Research Council 2008), area of watershed (Bi et al. 2014) etc. Numerous researches are done to study the effects of LULC on hydrological regime (Table 2.1).

2.3.2 Impact of Climate Change on Hydrology

Climate change has weakened the stationarity principle, a basic concept in water resource engineering which says that future hydrological events will fluctuate within the past variability (Milly et al. 2008). This is a result of human intervention in the natural processes of earth leading to changes in the means and extremes of evapotranspiration, precipitation and runoff. Studies are done to understand the trend of extreme rainfall events using various statistical methods (Babar and Ramesh 2014).

Commonly used climate change scenarios were those which are in the Special Report on Emission Scenarios (SRES) of Intergovernmental Panel on Climate Change (IPCC) Fourth Assessment Report (AR4) (IPCC 2007) (Ficklin et al. 2009; Tu 2009; Yoshimura et al. 2009; Praskievicz and Chang 2011). The Fifth Assessment Report (AR5) of the IPCC brought in new scenarios called representative concentration pathways (RCPs). Table 2.2 shows the studies conducted to ascertain the effects of climate change on various hydrologic parameters.

Table 2.1 Studies on impacts of land use land cover on hydrology

Author(s)	Study area	Study period	Hydrological model	DEM/Satellite	Key conclusions	Parameter
Babar and Ramesh (2015)	Netravathi River Basin, India	2001–2009	SWAT and RCRM	ASTER/IRS 1D LISS-3, Landsat	Sensitive parameters for the SWAT model were ascertained. RCRM which requires few input parameters can predict stream flows acceptably	Surface runoff
Sinha and Eldho (2018)	Netravathi river basin, India	1972, 1979, 1991, 2000, 2012 and 2030	SWAT	ASTER/Landsat	An increase in sediment yield and streamflow was observed due to the increase in urban and agricultural area and a decrease in the forest, leading to changes in the hydrological regime	Streamflow and sediment yield

(continued)

Table 2.1 (continued)

Author(s)	Study area	Study period	Hydrological model	DEM/Satellite	Key conclusions	Parameter
Zhu and Li (2014)	Little River Watershed, Tennesse	1984–2010	SWAT	National Elevation Dataset DEM, -Landsat	A 3% increase in streamflow was observed in whole watershed. 34.6% and 10% reduction in sediment and nutrient was respectively observed from 1984 to 2010, related to the reduction in agricultural land	Streamflow, Sediment yield, Total nitrogen, Total phosphorous
Petchprayoon et al. (2010)	Yom watershed, Thailand	1990–2006	MIKE 11	Landsat	Due to urbanization, the discharge rate was high at its downstream than that of other areas	Peak river discharge
Gyamfi et al. (2016)	Olifants Basin, South Africa	2000–2013	SWAT	SRTM/ Landsat	An increase of 46.97% in surface runoff was observed. Urbanization was identified as the major contributor to increases in surface runoff, water yield and evapotranspiration	Surface runoff, Lateral flow, Water yield, Groundwater Lateral flow, ET, Groundwater

Table 2.2 Studies on impacts of climate change on hydrology

Author(s)	Study area	Study period	Hydrological model	GCM/RCM	Key conclusions	Scenario	Parameter
Narsimlu et al. (2013)	Upper Sind River Basin, India	1992–2098	SWAT	PRECIS RCM	A significant increase in the runoff and baseflow is predicted. An increase in streamflow during monsoon season and a decrease in the offseason is also predicted	IPCC A1B Scenarios	Surface runoff
Ficklin et al. (2009)	San Joaquin River watershed, US	1992–2005	SWAT	–	The 50 year averaged outcomes show that the basin is highly sensitive to future climate changes	IPCC A1F1 and B1	Evapotranspiration, water yield, streamflow

(continued)

Table 2.2 (continued)

Author(s)	Study area	Study period	Hydrological model	GCM/RCM	Key conclusions	Scenario	Parameter
Treesa et al. (2017)	Wainganga river basin, India	1971–2040	VIC	CanESM2, IPSL-CM5A-MR, MIROC-ESM, ACCESS1-0, GISS-E2-R and GFDL- ESM2M	Streamflow did not show significant increase in the monsoon season but in the non-monsoon season	RCP 4.5	Streamflow
Soro et al. (2017)	Bandama Basin, West Africa	1986–2085	GR2M model	HadGEM2-ES model under	The results of the simulated impacts for RCP 4.5 and 8.5 were very different. The study highlights that there are huge uncertainties associated with impacts studies done with models	RCP 4.5 and RCP 8.5	Surface water and groundwater
Rwigi et al. (2016)	Sondu River basin, Kenya	1961–2050	SWAT	PRECIS	The study predicts overall increase in water yields with wetter dry seasons and drier wet seasons in the future	IPCC A2	Water yields

2.3.3 *Combined Effects of Climate Change and LULC Change*

Hydrological models simulate the impacts of climate and human activities on the hydrological regime when the hydro-meteorological data are fed to them. These can be used for finding the individual impacts of the same on streamflow. This is done by simulating the streamflow by assuming one of them constant and changing the other (Dey and Mishra 2017). SWAT has been extensively used for studying the effects of changing climate and land use land cover on streamflow (Dixon and Earls 2012). SWAT is equally popular in water quality studies (Glavan et al. 2015). Models like AVGWLF (Tu 2009), SWIM (Krysanova et al. 2015), PRMS (Legesse et al. 2003), MIKESHE (Wang et al. 2013), GR2M (Ahn and Merwade 2014), GBHM (Ma et al. 2010) etc. for same purposes. Studies considering both climate change and land use effects are shown in Table 2.3.

2.4 Model Comparison Studies

Tegegne et al. (2017) compared two conceptual models (GR4J, IHACRES) and a semi-distributed model (SWAT) in four watersheds of Ethiopia. The results could not conclude that any specific model was better than the other in all the watersheds. The conceptual models performed better in smaller watersheds than in largest watershed while SWAT performed better in largest watershed. Depending on the watershed, the model performance varied. The study further combined the models using ANN (Artificial Neural Network) and found reduction in RMSE (Root Mean Square Error) values. Coupling semi-distributed models with Artificial Neural Network (ANN) can improve the prediction of daily streamflow (Noori and Kalin 2016). An approach to integrate different models can help in better understanding the response of watersheds to LULCC and climate change (Kundu et al. 2017).

Krysanova and Hattermann (2017) did a comparison of impacts of climate change on 12 basins distributed in important geographical zones. The hydrological models used were ECOMAG, HBV, HYMOD, HYPE, mHM, SWAT, SWIM, VIC and WaterGAP3. The climate scenarios were simulated using HadGEM2-ES, IPSL-CM5A-LR, MIROC-ESM-CHEM, GFDL-ESM2M and NorESM1-M considering RCP2.6, RCP4.5, RCP6.0 and RCP8.5. The study concludes that the uncertainty associated with projections depend largely on selected GCM, followed by the selected RCP and hydrological model selected is the least contributor.

Table 2.3 Studies on impacts of land use land cover and climate change on hydrology

Author(s)	Study area	Study period	Model	Results	Parameter	Emission scenario	GCM/RCM
Chawla and Mujumdar (2015)	Upper Ganga basin (UGB), India	1971 to 2100	VIC model	Streamflow is extremely sensitive to modifications in the urban stretch while moderately sensitive to modifications in croplands. Climate change has a higher impact on streamflow than LULCC	Streamflow	RCP 4.5, 8.5	ACC, CNR, CCS, GFD, MPI and NOR
Ahn and Merwade (2014)	Indiana, New York, Arizona and Georgia area	1950–2010	GR2M	The impact of human influence on streamflow is greater than that of climate impact	Streamflow	–	–

(continued)

Table 2.3 (continued)

Author(s)	Study area	Study period	Model	Results	Parameter	Emission scenario	GCM/RCM
Kundu et al. (2017)	Elbow River watershed in southern Alberta, Canada	1961–2070	MIKE SHE/ MIKE 11	From the seasonal assessment it was understood that the LULC and climate change scenario gave higher streamflow in spring. The study emphasizes on the fact that based on the direction and magnitude of impact LULCC and climate change can offset or magnify each other's influence on the hydrological regime	Streamflow	A1B, B2	NCARPCM and CGCM2
Kim et al. (2013)	Hoeya River Basin, Korea	2020–2050	SWAT	The effect of LULC is less than that of climate change	Streamflow	RCP 4.5, 8.5	HadGEM3-RA
Mango et al. (2011)	Mara River Basin, Kenya	2002–2099	SWAT	Large variations in runoff was seen from small variations in precipitation and land use land cover	Streamflow	A1B	Average of 21 GCMs in the MMD (multi-model data set)

2.5 Discussion

This review deals with the studies that use hydrological modelling approach at basin scale to evaluate the impacts of land use land cover and climate change on hydrology. A brief note of model comparison studies is presented to detect the uncertainties associated with modelling. Based on the review it was observed that the model performance was dependent on the characteristics of the basin. Also, the impacts of land use land cover and climate change may magnify or diminish each other's effect on hydrology based on the catchment characteristics. Studies predicting future impacts of land use land cover and climate change on hydrology state significant changes in the hydrological parameters in the future. These findings demonstrate possible future water scarcity and related conflicts under the increasing urban development and extreme climatic conditions.

2.6 Conclusions

There is a huge usage of hydrological models in the study of impacts of land use land cover and climate change on water quality and quantity. Although there are uncertainties associated with the modelling process, it can be useful in understanding the complex interaction of various components and gives possible ranges of impacts in the coming years. There is a necessity of more studies evaluating the effects of climate change and land use land cover in many basins, as their interactive impacts are likely to occur and are not well understood. There are several gaps in area of hydrological modelling that could improve the accuracy of the outputs. Improvements in the downscaling methods, choice of models, quality of input data can improve the efficiency of modelling exercise. Quantification of uncertainties associated with the modelling results could improve the reliability of modelled future projections.

In the land use land cover change studies, the sensors on the landsat series of satellites are the popular ones because of the high temporal resolution and free availability of data. And, SWAT is the most popular among the hydrological models because it is less data intensive but gives reasonably accurate results. Based on the results of various modelling studies, it can be concluded that, depending on the characteristics of watershed and scenario assumptions, the combined effects of climate change and LULCC may ameliorate or deteriorate each other's influence. These effects may vary with the season or land use classes involved in change. Hydrological modelling integrating future climate change and LULCC scenarios can be an effective tool in planning future water resource management strategies.

References

Ahn K-H, Merwade V (2014) Quantifying the relative impact of climate and human activities on streamflow. J Hydrol 515:257–266
Babar S, Ramesh H (2014) Analysis of extreme rainfall events over Nethravathi basin. ISH J Hydraul Eng 20(2):212–221
Babar S, Ramesh H (2015) Streamflow response to land use—land cover change over the Nethravathi River Basin, India. J Hydrol Eng 20(10):1–11
Bi C, Bi H, Sun G, Chang Y, Gao L (2014) Scale effects and variability of forest—water yield relationships on the Loess Plateau, China. For Chron 90(2):184–191
Bormann H, Breuer L, Gräff T, Huisman JA, Croke B (2009) Assessing the impact of land use change on hydrology by ensemble modelling (LUCHEM) IV: Model sensitivity to data aggregation and spatial (re-) distribution. Adv Water Resour 32(2):171–192
Chawla I, Mujumdar PP (2015) Isolating the impacts of land use and climate change on streamflow. Hydrol Earth Syst Sci 19(8):3633–3651
Cristina L, Dias P, Macedo MN, Heil M, Coe MT, Neill C (2015) Regional studies effects of land cover change on evapotranspiration and streamflow of small catchments in the Upper Xingu River Basin, Central Brazil. J Hydrol 4:108–122
Dey P, Mishra A (2017) Separating the impacts of climate change and human activities on streamflow: a review of methodologies and critical assumptions 548:278–290
Dixon B, Earls J (2012) Effects of urbanization on streamflow using SWAT with real and simulated meteorological data. Appl Geogr 35(1–2):174–190
Dwarakish GS, Ganasri BP (2015) Impact of land use change on hydrological systems: a review of current modeling approaches. Cogent Geosci 1(1115691)
Fan M, Shibata H (2015) Simulation of watershed hydrology and stream water quality under land use and climate change scenarios in Teshio River watershed, northern Japan. Ecol Indicators J 50:79–89
FAO (2008) Forests and Water: a thematic study prepared in the framework of the global forest resources assessment 2005. FAO Forestry Paper 155
Ficklin DL, Luo Y, Luedeling E, Zhang M (2009) Climate change sensitivity assessment of a highly agricultural watershed using SWAT. J Hydrol 374(1–2):16–29
Glavan M, Ceglar A, Pintar M (2015) Assessing the impacts of climate change on water quantity and quality modelling in small Slovenian Mediterranean catchment - lesson for policy and decision makers. Hydrol Process 29(14):3124–3144
Gyamfi C, Ndambuki JM, Salim RW (2016) Hydrological responses to land use/cover changes in the Olifants Basin, South Africa. Water (switzerland) 8(12):1–16
Hundecha Y, Bárdossy A (2004) Modeling of the effect of land use changes on the runoff generation of a river basin through parameter regionalization of a watershed model. J Hydrol 292:281–295
Kim J, Choi J, Choi C, Park S (2013) Impacts of changes in climate and land use/land cover under IPCC RCP scenarios on stream flow in the Hoeya River Basin, Korea. Sci Total Environ 453:181–195
Krysanova V, Hattermann FF (2017) Intercomparison of climate change impacts in 12 large river basins: overview of methods and summary of results. Clim Change 141(3):363–379
Krysanova V, Hattermann F, Huang S, Hesse C, Vetter T, Liersch S, Koch H, Kundzewicz ZW (2015) Modelling climate and land-use change impacts with SWIM: lessons learnt from multiple applications. Hydrol Sci J 60(4):606–635
Kundu S, Khare D, Mondal A (2017) An integrated modelling system to predict hydrological processes under climate and land-use/cover change scenarios. Ecol Eng 105(10):1–23
Legesse D, Vallet-Coulomb C, Gasse F (2003) Hydrological response of a catchment to climate and land use changes in Tropical Africa: case study south central Ethiopia. J Hydrol 275:67–85
Leopold LB, Dunne T (1978) Water in environmental planning. W.H. Freeman
Li Z, Liu W-Z, Zhang X-C, Zheng F-L (2009) Impacts of land use change and climate variability on hydrology in an agricultural catchment on the Loess Plateau of China. J Hydrol 377(1–2):35–42

Lin Y, Verburg PH, Chang C, Chen H, Chen M (2009) Landscape and urban planning developing and comparing optimal and empirical land-use models for the development of an urbanized watershed forest in Taiwan. Landsc Urban Plan J 92:242–254
Ma H, Yang D, Tan SK, Gao B, Hu Q (2010) Impact of climate variability and human activity on streamflow decrease in the Miyun Reservoir catchment. J Hydrol 389(3–4):317–324
Mango LM, Melesse AM, McClain ME, Gann D, Setegn SG (2011) Land use and climate change impacts on the hydrology of the upper Mara River Basin, Kenya: Results of a modeling study to support better resource management. Hydrol Earth Syst Sci 15(7):2245–2258
Milly PCD, Betancourt J, Falkenmark M, Hirsch RM, Kundzewicz ZW, Lettenmaier DP, Stouffer RJ (2008) Climate change: stationarity is dead: whither water management? Science 319:573–574
Mudbhatkal A, Amai M (2017) Regional climate trends and topographic influence over the Western Ghat catchments of India. Int J Climatol 38(5):2265–2279
Mudbhatkal A, Raikar RV, Venkatesh B, Mahesha A (2017) Impacts of climate change on varied river-flow regimes of Southern India. J Hydrol Eng 22(9):1–13
Narsimlu B, Gosain AK, Chahar BR (2013) Assessment of future climate change impacts on water resources of Upper Sind River Basin, India Using SWAT Model. Water Resour Manage 27(10):3647–3662
National Research Council (2008) Hydrologic effects of a changing forest landscape
Noori N, Kalin L (2016) Coupling SWAT and ANN models for enhanced daily streamflow prediction. J Hydrol 533:141–151
Petchprayoon P, Blanken PD, Ekkawatpanit C, Hussein K (2010) Hydrological impacts of land use/land cover change in a large river basin in central-northern Thailand. Int J Climatol 30(13):1917–1930
Piao S, Ciais P, Huang Y, Shen Z, Peng S, Li J, Zhou L, Liu H, Ma Y, Ding Y, Friedlingstein P, Liu C, Tan K, Yu Y, Zhang T, Fang J (2010) The impacts of climate change on water resources and agriculture in China. Nature 467(7311):43–51
Praskievicz S, Chang H (2009) A review of hydrological modelling of basin-scale climate change and urban development impacts. Prog Phys Geogr 33(5):650–671
Praskievicz S, Chang H (2011) Impacts of climate change and urban development on water resources in the Tualatin River Basin, Oregon. Ann Assoc Am Geogr 101(2):249–271
Rodríguez-Blanco M, Arias R, Taboada-Castro M, Nunes J, Keizer J, Taboada-Castro M (2016) Potential impact of climate change on suspended sediment yield in NW Spain: a case study on the Corbeira catchment. Water 8(10):444
Rwigi SK, Muthama JN, Opere AO (2016) Simulated impacts of climate change on surface water yields over the Sondu Basin in Kenya. Int J Innov Educ Res 4(8):160–172
Sinha RK, Eldho TI (2018) Effects of historical and projected land use/cover change on runoff and sediment yield in the Netravati river basin, Western Ghats, India. Environ Earth Sci 77(111):1–19
Soro G, Yao A, Kouame Y, Bi T (2017) Climate change and its impacts on water resources in the Bandama Basin, Côte D'ivoire. Hydrology 4(1):18
Tegegne G, Park DK, Kim YO (2017) Comparison of hydrological models for the assessment of water resources in a data-scarce region, the Upper Blue Nile River Basin. J Hydrol: Reg Stud 14:49–66
Treesa A, Das J, Umamahesh NV (2017) Assessment of impact of climate change on streamflows using VIC model. Euro Water 59(2013):61–68
Tu J (2009) Combined impact of climate and land use changes on streamflow and water quality in eastern Massachusetts, USA. J Hydrol 379(3–4):268–283
Wang S, Zhang ZR, McVicar T, Guo J, Tang Y, Yao A (2013) Isolating the impacts of climate change and land use change on decadal streamflow variation: assessing three complementary approaches. J Hydrol 507:63–74
Weatherhead EK, Howden NJK (2009) The relationship between land use and surface water resources in the UK. Land Use Policy 26(SUPPL. 1):S243–S250

Yoshimura C, Zhou M, Kiem AS, Fukami K, Prasantha HHA, Ishidaira H, Takeuchi K (2009) 2020s scenario analysis of nutrient load in the Mekong River Basin using a distributed hydrological model. Sci Total Environ 407(20):5356–5366
Zhou F, Xu Y, Chen Y, Xu C, Gao Y, Du J (2013) Hydrological response to urbanization at different spatio-temporal scales simulated by coupling of CLUE-S and the SWAT model in the Yangtze River Delta region. J Hydrol 485:113–125
Zhu C, Li Y (2014) Long-term hydrological impacts of land use / land cover change from 1984 to 2010 in the Little River Watershed, Tennessee. Int Soil Water Conserv Res 2(2):11–21

Chapter 3
Water Resource Management for Coal-Based Thermal Power Plant

S. A. Nihalani, Y. D. Mishra, and A. R. Meeruty

Abstract Water and energy are elementary requirements for well-being and prosperity of humans. They are mutually dependent on each other. Energy production involves a huge quantity of water and in turn, water structures need a huge amount of energy. This dependency is termed as 'water-energy nexus'. As a result of population and economic growth, and higher living standard, future demand for both water and energy is anticipated to upsurge, particularly in the emerging economies like India. The majority of energy-linked water usage is for power generation, which is led by water-concentrated electricity generation from coal, natural gas, and nuclear materials. Majority of the water utilised at thermal power plants is for cooling purpose in addition to ash handling, boiler feed water, flue gas desulphurisation and for other applications in coal-fired power plants. Governing authorities shall impose limits on freshwater usage by power plants by framing regulatory norms. The present stress on freshwater supplies and increasing demand is compelling the power generating units to search for alternative sources or supplementary sources of water. Utilising water from such sources will help conserve fresh water for other uses, such as drinking and agriculture. The current paper discusses water resource management for coal-based thermal power plants. If further evaluates various alternatives or measures for water conservation in thermal power plants.

Keywords Air cooled condenser · Evaporation loss · Makeup · DM plant · Zero liquid discharge

S. A. Nihalani (✉) · A. R. Meeruty
Civil Engineering Department, PIET, Parul University, Vadodara, Gujarat, India
e-mail: seema.nihalani@paruluniveristy.ac.in

Y. D. Mishra
Mechanical Department, L&T-S&L, Vadodara, Gujarat, India

R. Jha et al. (eds.), *Water Resources Management and Reservoir Operation*, Water Science and Technology Library 107,
https://doi.org/10.1007/978-3-030-79400-2_3

3.1 Introduction

Water is one of the unavoidable input needed for sustainable growth of the industrial sector of any country. As the need for power increases, the demand for freshwater supply for power generation will also grow. During past years, there is a decrease in the availability of freshwater. Also, there is a rising realization in power plant managers to preserve fresh water and decrease the use of freshwater by employing various water conservation techniques including treatment, recycle and reuse. In addition, the need to leave water in receiving stream to meet environmental regulations will further govern the future provision of freshwater supply.

Ministry of Environment and Forest has taken stringent steps to control environment and water pollution. MOEF has recently issued notification for consumption of water usage and emission standards within thermal power plants. Based on the notification issued by MOEF, all the existing thermal power plants have to reduce specific water consumption up to 3.5 m^3/MWh. All the new plants that shall be installed after 1st January 2017 are required to meet specific water consumption up to a maximum of 2.5 m^3/MWh and must achieve zero liquid discharge (ZLD). This new environment norm generates the need for every power generation company to critically evaluate the water resource management of coal-based power plant. The first step in water resource management is to evaluate the water requirement of various consumers in power plant and understand the quality of water required at various sources. This paper will evaluate the possible alternatives to reduce the fresh makeup water requirement and show ways to achieve zero liquid discharge. This paper will also discuss the adoption of air-cooled condenser technology in place of the conventional wet cooling system to meet the new environment norms.

3.2 Water Resource Management in Thermal Power Plants

Water requirement in power generation is regulated by a number of factors like raw water quality, condenser cooling system, coal quality, utilization of ash, disposal of ash, wastewater handling etc. Based on water quality and end-user concerns, the water utility in power plants can be divided into following types:

1. Ash Handling Water: water utilized to convert dry ash to slurry form for disposal.
2. Cooling Tower water: Water utilized for cooling towers.
3. Makeup Water: Water utilized for compensating evaporation losses in the cooling tower.
4. Demineralised Water: Water utilized for generating steam in boilers.
5. Drinking Water: Water used for human consumption purpose.
6. Utility Water: Water used for various services like fire-fighting, dust suppression, gardening, and sanitation purposes etc.

3.2.1 Cooling Water System Water Requirements

Cooling water system mainly uses water to condense the steam coming out from LP turbine exhaust. The heat of condensed steam will again be rejected to the atmosphere with the help of a wet cooling tower. In a closed cooling water system, cooling tower rejects the heat of hot water coming out from the condenser by evaporating some amount of water in the atmosphere. The process of evaporation inside cooling tower increases the total dissolved solid levels of circulating water which in turn increases scaling and corrosion problems in the system. To maintain dissolved solids levels in the closed cooling water system, some amount of water needs to be drained out from the system by making up fresh water to the system. The limit of dissolved solids levels in the cooling water system is dictated by a cycle of concentration. The cycle of concentration (COC) is how many times dissolved solids in water is getting concentrated without increasing scaling/corrosion issues and the loss of water to maintain COC of the system is termed as CW blow-down. In a closed cooling water system, water is required to be makeup to compensate for evaporation loss, blow-down loss and drift losses. Increasing COC of the system reduces the blow-down loss from the system however; it will impact circulating cooling water quality and affect the performance of the circulating cooling water system. Generally, river water based power plants are designed to maintain COC as 5 and coastal power plants using closed cooling water system are designed to maintain COC in range of 1.2 to 1.3.

3.2.2 Ash Handling System Water Requirements

Ash is generated due to the burning of coal inside the boiler, which needs to be disposed of off to ash dyke. There are two types of ash generated from the boiler. Ash collected at bottom of boiler and economizer is termed as Bottom ash and ash collected at bottom of ESP is termed as Fly ash. Generally, bottom ash and fly ash are generated in the ratio of 20% and 80% of the total ash generation during the burning of coal. In ash handling system, water is required mainly to prepare ash slurry in specific concentration for evacuating bottom ash and fly ash to ash dyke with the help of slurry pumps. Water requirements in ash handling plant mainly depend on the type of ash handling system and quantity of ash available in the coal. For bottom ash disposal, mostly the practice is to use lean slurry disposal system in which 20–22% concentrated slurry is made with the help of water and transferred to ash dyke with the help of bottom ash slurry disposal system. However, a new trend of the dry bottom ash handling system is finding a place in some of the recent tenders. For fly ash disposal, industry proven practice is to use either of the following systems:

- The lean slurry disposal system
- High concentrated slurry disposal system (HCSD)
- Dry Fly ash disposal system.

High concentrated slurry disposal system (HCSD) uses disposal of 50–55% concentrated fly ash slurry through HCSD pumps. Hence, water requirements of HCSD system are less as compared to the lean slurry fly ash disposal system. Further, due to MOEF norms for 100% utilization of fly ash from the 4th year of plant operation, all the power plants since this notification have been designed to dispose fly ash in wet mode (through Lean slurry or HCSD system) during initial period of plant operation and then dispose of via dry fly ash disposal system. Power plants have also been equipped with ash water recirculation system to recover clear water from ash dyke area. With the installation of ash water recirculation system, approximately 70% water can be recovered and utilized in ash handling system for slurry preparation after 1 year of plant operation.

With the help of water recovery from ash dyke after one year and 100% dry fly ash utilization after 4 years of plant operation, water requirements in the ash handling plant are reduced to a considerable extent. However, during initial periods of plant operation, a considerable amount of water requirement is envisaged for wet bottom ash and fly ash disposal system with no recovery from ash dyke. For a typical 2 × 660 MW plant capacities with Indian coal quality that consists of high ash content, the total makeup water requirement for ash handling plant is estimated in the following range.

Table 3.1 shows that ash handling plant consumes 1.3–1.5 m^3 of water for production of 1 MWh electricity. For the plant using lean slurry disposal system for bottom ash and fly ash disposal, above water requirement is reduced to 500–600 m^3/h after the first year of plant operation when recovery of water from ash dyke area is available. However, for a plant using the HCSD system for fly ash disposal, water requirement remains in the range of 700 m^3/h because no water could be recovered from fly ash HCSD system. However, during the fly ash utilization period, the water requirement will be in the range of 220–250 m^3/h approximately.

Total makeup water requirement of ash handling plant during the initial period of plant operation is generally being met by cooling tower blowdown water and remaining by raw water. During the ash water recovery scenario, water requirement can be met by cooling tower blowdown water and no additional raw water is needed. However, during less water requirement in case of fly ash utilization after the 4th year of plant operation, cooling tower blowdown water could not be fully utilized in ash handling plant for slurry making and this creates a situation of excess availability of cooling tower blowdown water. The scenario of existing power plants using dry fly ash utilization and recovery water from ash dyke creates concern for power Generation Company on how zero liquid discharge could be achieved.

Table 3.1 Water requirement for ash handling

Description	Bottom ash and Fly ash disposal the system in a lean slurry mode with No ash water recovery	Bottom ash disposal in lean slurry mode and Fly ash disposal in HCSD mode with No ash water recovery
Makeup water requirement	1800–2000 m^3/h	1300–1400 m^3/h

3.2.3 Coal Handling System Water Requirements

In a coal-based thermal power plant, coal is used as working fuel in the boiler for generating steam from water. The required amount of coal needs to be met from nearby coal mines through conveyer/railway wagon/trucks. In the process of making and transporting suitable size of coal in the boiler from Coal Storage area, water is required to spray during unloading and transportation to avoid the emission of coal particles in the atmosphere and to spray water at coal storage area to avoid spontaneous ignition of coal. Water Requirements for coal handling plant varies from plant to plant based on the location of coal storage, numbers of coal Stockpiles, numbers of transfer points. For typical 2 × 660 MW unit, water requirement for coal handling plant is estimated in the range of 100–150 m^3/h.

3.2.4 Demineralized Water System Water Requirements

Demineralized water is mainly used as power cycle makeup in condenser hot well to compensate blowdown and other steam vent losses that occur in case of subcritical power plants. In supercritical power plants, cycle makeup is only required to compensate for the steam vent losses and other losses due to leakages. Power cycle makeup for supercritical power plants is generally considered as 0.5–1% of BMCR steam flow. Further consumers of demineralized water in power plant are condensate polishing units, demineralized water generation plant, closed auxiliary cooling water system. Condensate polishing plant is installed in all supercritical power plants to maintain the stringent quality of demineralized water in the power cycle as well as to reduce power cycle makeup water requirement. Condensate polishing plant is placed at the condensate extraction pump discharge to treat the condensate coming from the condenser. In the condensate polishing plant, demineralized water is required to regenerate the resins of condensate polishing units. DM water is also used as makeup for the closed auxiliary cooling water system to compensate for water losses due to leakages. For Typical 2 × 660 MW supercritical power plant, total demineralized water requirement is estimated in the range of 50–60 m^3/h.

3.3 Water Balance for 2 × 660 MW Coal-Based Power Plant

The first step in the development of optimized water resource management for the coal-based power plant is to review the present scenarios of water balance in the plant. For a review of present water resource management, the water balance of 2 × 660 MW coal-based supercritical power plant has been prepared considering all modern practices followed by recent power plants. Following design considerations are adopted to develop the typical water balance of 2 × 660 MW plant:

- River water is considered a source of fresh raw water.
- Water cooled condenser is considered for circulating cooling water system and the cycle of concentration for the system is considered as 5.
- Evaporation loss and drift loss inside the wet cooling tower is considered as 1.6% and 0.005% of cooling tower flow respectively.
- Cooling tower blowdown is waste having high dissolved solids. However, this waste will be utilized directly without treatment for ash slurry preparation in ash handling plant.
- Total demineralized water requirement is considered as 1.5% of BMCR flow considering regeneration water requirements and Conventional Ion exchange resin based demineralization (DM) plant is considered to generate the DM quality water.
- Demineralization and condensate polishing plant waste is generated during regeneration of ion exchange resin. This waste is having high dissolved solids and will be utilized in ash handling and coal handling plant after neutralization of waste.
- Pre-treatment plant sludge generated at bottom of clarifiers will further be treated using thickener and dewatering centrifuge for generation of cake in semi-dry form.
- Floor wash along with oily waste generated will be treated in the clarifier and oily water separator to recover clear water. Clearwater will then be utilized to meet ash handling and coal handling plant water requirements.
- Lean slurry disposal system is considered for bottom ash disposal and high concentrated slurry disposal system (HCSD) is considered for fly ash disposal during initial operation of the plant.
- Approximately 70% of water from ash dyke is considered to be recovered after 1 year of plant operation and 100% fly ash will be utilized after 4th years of plant operation.
- Water requirement for future flue gas desulfurization (FGD) units is not considered while reviewing the present scenario of water balance.

Based on the above design consideration, net raw water makeup requirement for typical 2 × 660 MW coal-based supercritical plant has been estimated as 4230 m^3/h which corresponds to specific water consumption of around 3.2 m^3/MWh. This water requirement is established considering no recovery water received from an ash pond during initial years of plant operation (Table 3.2).

The specific water consumption will further be reduced during ash water recovery scenario as well as 100% dry fly ash utilization period. However, during 100% dry fly ash utilization period there will be excess cooling tower blowdown in the range of 220–250 m^3/h within the plant and zero liquid discharge could not be achieved. Further, these values are estimated considering river water based plant and COC of 5. For a plant which receives bore well water and seawater as raw water above specific water consumption values increases significantly. The specific water consumption values for current scenarios of river water and seawater-based plant using closed cooling water system are shown in Table 3.3.

Table 3.2 Consumptive water requirement of 2 × 660 MW power plant

S. No.	Systems	Water consumption (m^3/h)	Specific water consumption (m^3/MWh)
1	Makeup water requirement in the condenser cooling water system	3000	2.27
2	Total makeup water requirement for ash handling plant	1400	1.06
3	Total makeup water requirement for coal handling plant	150	0.11
4	DM water requirements	60	0.05
5	Miscellaneous service water & potable water requirements including HVAC, pump sealing	250	0.19
	Total water requirements	4860	3.68
6	CW blowdown water generated	600	
7	Other waste generated including DM plant waste, CPU waste, floor washing waste, and oily waste	30	
	Total waste collected	630	
9	Reduction in ash handling plant makeup water requirement due to the utilization of plant waste	480	
10	Reduction in coal handling plant makeup water requirement due to the utilization of plant waste	150	
11	Net raw water makeup requirement	4230	3.2

3.4 Analysis of New Environment Norms and Its Impacts

Study of new environment norms issued by MOEF reveal following facts with respect to coal-based power plants:

- Once through cooling system is now discarded for all existing power plant as well as for upcoming new power plants. Environment rules mandate to install a closed cooling water system for all existing once-through power plant. Further, rules also mandate to achieve specific water consumption of 3.5 m^3/MWh for all existing once-through power plant within 2 years.

Table 3.3 Water consumption for river and sea water based plant

S. No.	Description	Without ash water recovery	With ash water recovery	With ash water recovery and 100% fly ash utilization
A	*Seawater-based plant considering COC of 1.3*			
1	Specific water consumption	8.7 m^3/MWh	8.7 m^3/MWh	8.7 m^3/MWh
2	With Ash water recovery and 100% fly ash utilization	7300 m^3/h	7800 m^3/h	8400 m^3/h
3	Is zero liquid discharge achieved	No	No	No
B	*River water-based plant considering COC of 5*			
1	Specific water consumption	3.2 m^3/MWh	2.65 m^3/MWh	2.5 m^3/MWh
2	Excess CW blowdown water available in plant	0 m^3/h	0 m^3/h	220–250 m^3/h
3	Is zero liquid discharge achieved	Yes	Yes	No

- All new coal-based plants to be installed after 1st January 2017 should reduce specific water consumption value to 2.5 m^3/MWh and should achieve zero liquid discharge.
- All existing and new coal-based power plants have to meet the stringent requirements of SO_2 and NO_x emissions. To meet the new emission standards, most of the power plant has to install Flue gas desulfurization units.

From above, it is observed that all the upcoming power plants irrespective of water sources available (i.e. Sea water /river water /bore well) have to install closed cooling system and flue gas desulfurization units. Further, flue gas desulfurization system also uses water for removal of SO_2 and other impurities from flue gases. There are various technologies available for SO_2 removal from the flue gas stream. Among various technologies including sea water based FGD system, wet limestone FGD system is considered in this assessment for large coal-based power plants. Makeup water requirement for Wet limestone FGD system has been estimated in the range of 220–250 m^3/h based on supplier's inputs for 2 × 660 MW units.

In view of above scenario, it is observed that new environment norms could not be achieved with the help of present water resource management adopted in coal-based power plants using sea water and plant where higher COC could not be possible. The implementation of new environment norms must need to evaluate various alternatives for further optimization of water requirements and achievement of zero liquid discharge in coal-based power plants.

3.5 Adoption of Dry Cooling System (Air Cooled Condenser)

In this alternative, water resource management using an air-cooled condenser is evaluated with no change in ash handling system operation. In air-cooled condenser heat is rejected to the atmosphere by sensible cooling only. It does not involve evaporative cooling. Hence makeup water requirement for condenser cooling reduces to zero. The water resource management for a plant using an air-cooled condenser with fly ash HCSD system and plant using wet cooling tower with fly ash HCSD system can be correlated as Table 3.4.

Above table shows that specific water consumption value of this alternative is almost 50% lower than the wet cooling system. With this alternative, new environment norms of specific water consumption, as well as zero liquid discharge, could easily be achieved in all operating scenarios of plant irrespective of the quality of water.

Table 3.4 Consumptive water requirement of 2 × 660 MW power plant with WCC and ACC

S. No.	Systems	Water consumption (m^3/h)	Specific water consumption (m^3/MWh)	Water consumption (m^3/h)	Specific water consumption (m^3/MWh)
		Water-cooled condenser		Air-cooled condenser	
1	Makeup water requirement in the condenser cooling water system	3000	2.27	0	0
2	Makeup water requirement in the auxiliary cooling system			200	0.15
3	Total makeup water requirement for ash handling plant	1400	1.06	1400	1.06
4	Total makeup water requirement for coal handling plant	150	0.11	150	0.11

(continued)

Table 3.4 (continued)

S. No.	Systems	Water consumption (m^3/h)	Specific water consumption (m^3/MWh)	Water consumption (m^3/h)	Specific water consumption (m^3/MWh)
		Water-cooled condenser		Air-cooled condenser	
5	DM water requirements	60	0.05	60	0.05
6	Miscellaneous service water and potable water requirements including HVAC, pump sealing	250	0.19	250	0.19
7	FGD system makeup water requirements	300	0.22	300	0.22
8	Total water requirements	5160	3.9	2360	1.8
9	CW blowdown water generated	600		70	
10	Other waste generated including DM plant waste, CPU waste, floor washing waste, and oily waste	30		30	
11	Total waste collected	630		100	
12	Reduction in ash handling plant makeup water requirement due to the utilization of plant waste	480		100	
13	Reduction in coal handling plant makeup water requirement due to the utilization of plant waste	150		0	

(continued)

Table 3.4 (continued)

S. No.	Systems	Water consumption (m^3/h)	Specific water consumption (m^3/MWh)	Water consumption (m^3/h)	Specific water consumption (m^3/MWh)
		Water-cooled condenser		Air-cooled condenser	
14	Net raw water makeup requirement	4530	3.4	2260	1.7

3.6 Additional Water Conservation Techniques Used in TPP

The utility of tremendously large volumes of water in the thermal power plants demands that the thermal power plant companies undertake numerous water conservation measures. Apart from the use of an air-cooled condenser, following water conservation techniques may also help to achieve new environmental norms of water consumption for many power plants.

3.6.1 Adoption of 100% Utilization of Fly Ash in Dry Mode

Present water balance of river water based plant using HCSD system for fly ash disposal estimates specific water consumption in the range of 3.2 m^3/MWh during the initial period of plant operation. This meets the criteria of new environment norms specified for many existing power plants. These power plants could achieve specific water consumption of 3.5 m^3/MWh by utilizing 100% dry fly ash disposal in place of conventional lean slurry disposal system and using recovery water from ash dyke.

For river water based plants to be installed after January 2017, new environment norms of 2.5 m^3/MWh could be achieved using 100% fly ash disposal in dry mode after 1 year of plant operation when recovery of water also starts from ash dyke area. However, during the above scenario, there was surplus CW blowdown water in the range of 220–250 m^3/h which could now be utilized to meet the makeup water requirement of recently added FGD system. Further, a wet FGD system also generates waste in the range of 10–15 m^3/h for 2 × 660 MW capacity plant. As waste from FGD system is less, this can be disposed to ash dyke after the removal of suspended solids and heavy metals through a suitable treatment process. With above consideration, zero liquid discharge could be possible in the plant. This alternative helps to achieve new environment norms for river water based plant after 1 year of plant operation without a substantial change in present systems of the plant. However, for plants where 5 COC of water could not be achieved due to the quality of raw water (in case of bore well and seawater), it is not possible to achieve new environmental norms of 2.5 m^3/MWh and zero liquid discharge with the adoption of 100% dry fly ash utilization.

3.6.2 Increasing Cycle of Concentration for Circulating Cooling Water System

Makeup water requirement in the CW system is a summation of cooling tower evaporation loss, CW blowdown and cooling tower drift loss. Cooling tower evaporation loss and drift loss are a function of cooling tower performance and could not be reduced. CW blowdown loss is a function of the cycle of concentration being maintained in the CW system. The cycle of concentration for CW system will depend on the quality of makeup water. Generally, river /canal water having low dissolved solids may achieve a higher cycle of concentration in the range of 6 to 7. Most of the recent river water based plant has been designed to achieve 5 COC. Increasing cycle of concentration will reduce the amount of CW blowdown waste which in turn reduce net CW make-up water requirement. For typical 2 × 660 MW plant, increasing COC from 5 to 6 will reduce CW blowdown quantity in the range of 60–80 m^3/h.

3.6.3 Installation of Ash Water Recovery System from Ash Dyke

Ash dyke receives the ash slurry from bottom ash and fly ash disposal system. In Ash dyke, heavy particles of ash will get settled at the bottom. Over a period of time, clear water from the ash slurry will overflow from ash dyke area. To reuse clear water received from ash dyke area, recovery system will be installed. In Ash water recovery system, clear water overflow from ash dyke area will be passed through clarifiers for further clarification. Clarified water is then recycled/reused for ash slurry preparation. Generally, overflow from ash dyke area starts after one year of plant operation based on ash dyke volume. Installation of ash water recovery system recovers approx. 70% of water. Hence, Ash handling plant water requirement during ash water recovery scenario will reduce to 30% of water being supplied at initial plant operation.

3.6.4 Recycling of CW Blowdown to Other Systems

CW blowdown is waste generated from circulating cooling water system to maintain dissolved solids levels. CW blowdown can directly be reused for ash slurry preparation as well as for coal dust suppression system in power plant. Reusing of CW blowdown in ash handling plant and coal handling plant also reduces fresh makeup water requirements. However, dry fly ash utilization, as well as ash water recovery scenario, creates a situation of excess CW blowdown waste available within the plant. This excess CW blowdown water needs to be treated and recycled suitably to meet the zero liquid discharge condition.

3.7 Conclusion

New environment norms for water reduction and zero liquid discharge was reviewed along with existing plant water resource management and possible alternative options are discussed and compared to optimize the water requirements for existing as well as future power plants. The outcome of various alternative options discussed in this paper are summarized below:

- For Indian conditions, adopting dry cooling system shall result in a reduction of plant output by about 7% and similarly, the gross heat rate shall increase by about 7%. The capital cost of the plant per MW shall increase by about 10%.
- For future river water based plants, installation of wet limestone FGD system, HCSD system for fly ash disposal and achieving 100% utilization of dry fly ash from 1st year could achieve the specific water consumption of 2.5 m^3/MWh with zero liquid discharge after 1 year of plant operation when recovery of water from ash dyke starts.
- For existing river water based plants, installation of wet limestone FGD system and achievement of
- 100% dry fly ash utilization could meet the new environment norms.
- For seawater based once through the plant, Installation of the closed cooling system, as well as FGD system, is now necessity of new environment norms. Since seawater FGD plant requires a huge amount of seawater, wet limestone FGD system is the probable solution to reduce the FGD water requirements in seawater-based plants. Further, Installation of the wet cooling tower for seawater based plants also end up with a huge amount of excess CW blowdown water in the range of 7300–8400 m^3/h. Achievement of zero liquid discharge with such huge amount of excess blowdown does not seem practical with the help of available water treatment technologies in the market.
- Specific water consumption of 3.5 m^3/MWh could also not be achieved for existing seawater based plant with the installation of wet cooling towers. At present, the dry cooling system (air cooled condenser) is only probable alternatives to meet the new environment norms for seawater-based plants. However, installation of the air-cooled condenser to existing seawater based plants using a water-cooled condenser is challenging in terms of meeting rated plant output as well as retrofitting of the air-cooled condenser in place of the water-cooled condenser.
- For new as well as existing plants using bore well water which operates on low COC of 3–4, specific water consumption along with zero liquid discharge could be achieved by utilizing 100% fly ash and installing wastewater treatment plants for excess blowdown water. However, in some of the cases, costly thermal desalination technology may be required to achieve zero liquid discharge.
- Industry and Academia are working towards reducing the operating performance and capital cost gap between dry cooling and the wet cooling. With stringent environmental norms and scarcity of water favours use of the dry cooling system and would be the prominent technology in future.

Acknowledgements The study is part of Industry Defined Research Project titled "To Reduce Water Consumption and Optimize Efficiency of Thermal Power Plants by Using Air Cool Condensers" funded by Royal Academy of Engineering under Newton Bhabha Fund in collaboration of L&T Power with Parul University.

References

Notification issued by Ministry of Environment, Forest and Climate Change dated 7 December 2015

Water Use and Efficiency in Thermal Power Plants, Report by FICCI

Water balance diagram for 2 x 660 MW plant engineered by L&T-Sargent & Lundy Limited

Chapter 4
Evaluation of Reservoir Sedimentation Using Satellite Data—A Case Study

Beeram Satya Narayana Reddy and S. K. Pramada

Abstract India is an immense nation with high spatial and temporal inconsistency of precipitation. The end target is to tap the available water assets and to use the water as per the necessity. Numerous dams have already constructed to tap the available water resources. But the storage capacity of these reservoirs is gradually reducing due to sedimentation. To decide the useful existence of reservoir and to evaluate the sedimentation rate in the reservoir, it is fundamental duty to conduct the surveys intermittently. If the actual sedimentation process evaluated in the reservoirs, measures can propose ahead of time, and reservoir operational plan can arrange for typical usage of water. Some traditional procedures of measurement of reservoir sedimentation like hydrographic surveys and inflow-outflow techniques are expensive and also tedious. Remote sensing method turns into a helpful instrument for assessment of reservoir sedimentation because of its spatial, spectral and time-related features, which give succinct, monotonous and auspicious information concerning the area of water spread of the reservoir. Because of the deposition of sediments in a reservoir, the water spread at a height continuously alters. Presently by looking at the decreasing area of water spread with time, the sedimentation rate in a reservoir can be determined easily. In the current research, the sedimentation rate and volume ascertain within the Thenmala reservoir were carried out utilising satellite data. The data of IRS-P6 satellite with LISS-III sensor for 6 different date of pass was used to extract the area of water spread. The initial elevation-area-capacity curves and the reservoir levels on the six date of pass were acquired from the authorities of the dam. Utilising the cone formula, the capacity in zone of live storage is calculated. Because of silt deposition inside the reservoir, 44.063 (M Cu.m) storage capacity lost in a traverse of 30 years, i.e. 1985–2014.

Keywords Linear Imaging Self Scanning (LISS-III) · Normalised Difference Water Index (NDWI) · Area of water spread · Reservoir sedimentation · Satellite remote sensing

---

B. S. N. Reddy (✉) · S. K. Pramada
Civil Engineering, NIT, Calicut, Kerala, India
e-mail: satyanarayana_p170016ce@nitc.ac.in

R. Jha et al. (eds.), *Water Resources Management and Reservoir Operation*, Water Science and Technology Library 107,
https://doi.org/10.1007/978-3-030-79400-2_4

4.1 Introduction

Dam is designed across river aiming in irrigation, power generation, regulation and supervision of discharge. Nowadays many reservoirs are not performing their design operation because the sediment is occupying the operational capacity of storage (Ijamand Al-Mahamid 2012). Once the stream enters a storage reservoir, attributable to the small flow rate in the reservoir, they tend to the settlement of sediment. As settlement of silt happens bit-by-bit, operational capacity in the reservoir reduced which makes inefficient of performing actual purpose of impoundment of dam as time passes.

Due to the construction of dams on natural streams which forms a reservoir will subject to some extent of silt influx and accumulation, which in turn reduces the reservoir life. Periodical capacity surveys of reservoirs facilitate in judging the speed of sediment deposition and depletion in capacity of storage (Jeyakanthan and Sanjeevi 2013). Bathymetry and sediments spatial diffusion modelling, geographical information system (GIS) can be adapted for sedimentation studies. Many researchers had used remote sensing technology in the assessment of sediment accumulation (Goel et al. 2002; Jain et al. 2002; Pandey et al. 2016; Yeo et al. 2014; Narasayya et al. 2013). Usage of satellite image recently becomes economical and expedient to figure out the sediment deposition and to determine its dispersal and accumulation pattern (Narasayya et al. 2013).

Satellite image techniques offer information obtainment over an extended amount of duration which provides comprehensive, monotonous and timeous information about reservoir sediment deposition nature. Reservoir area of water spread at specific elevation may be acquired meticulously with satellite imagery. Curtailment if any, within the region of water spread at particular elevation is due to sediment deposition. Multi-date of the pass of satellite imagery related to the study area is required, from which area of water spread extracted will permit assessment of storage volume lost due to sediment deposition.

The present aim of the study is to assess the capacity loss in Thenmala reservoir, Kollam, Kerala, India due to sedimentation through satellite data, to update elevation-area-capacity curve and to estimate the loss of storage due to sediment deposition.

4.2 Description of Study Area

Thenmala dam was constructed in the year 1985 on the river Kallada in Kollam district in Kerala. Kallada River is a west-flowing river which originates from Kulathupuzha ranges. After flowing through 120 km in Kollam district, it falls into Ashtamudi Lake. The study area lies between latitudes 08° 49′ N to 09° 17′ N and longitudes 77° 16′ E to 76° 24′ E. It covers portions of Thiruvananthapuram, Kollam and Alleppey districts. The location of the dam is 08° 57′ N and 77° 4′ E. Kallada River forms by the confluence of three major tributaries of river viz., Kulathupuzha, Senthuri and

Kalthuruthi. The map of the study area is shown in Fig. 4.1. Area of watershed is 549 km^2. Main benefits of reservoir are irrigation and power generation. The salient features of the dam are shown in Table 4.1.

Fig. 4.1 Location of study area

Table 4.1 Salient features of Thenmala dam

Official name	Kallada dam
Location	Kollam District, Kerala
Coordinates	08° 57′ 00″ N 77° 04′ 20″ E
Purpose	Irrigation, hydro power
Status	Operational
Opening date	1986
Dam and spillways	
Height	85.35 m
Length	335 m
Spillways	3
Spillway type	Ogee with radial gate auxiliary-Labyrinth
Reservoir	
Catchment area	549 km^2
Surface area	23 km^2
Gross storage	504.92 Mm^3
Live storage	487.92 Mm^3
Dead storage	17 Mm^3
Storage between RL 88 m to FRL 115.82	423.953 Mm^3

Table 4.2 Satellite image collected from NRSC Hyderabad

Satellite	Sensor	Path	Row	Date of pass	Elevation in m
IRS-P6	LISS-III	100	67	22 February 2008	109.3
IRS-P6	LISS-III	100	67	20 January 2012	112.22
IRS-P6	LISS-III	100	67	05 October 2013	115.15
IRS-P6	LISS-III	100	67	16 December 2013	114.05
IRS-P6	LISS-III	100	67	26 February 2014	106.72
IRS-P6	LISS-III	100	67	04 January 2015	114.25

4.3 Data and Software Used

LISS-III sensor data of IRS-P6 satellite with a spatial resolution 23.5 m used for the analysis. The data was purchased from National Remote Sensing Centre (NRSC) Hyderabad. Data collected from NRSC Hyderabad of LISS-III was available in 4 bands of different wavelengths regions like Band-2 (Green), Band-3 (Red), Band-4 (NIR) and Band-5 (SWIR) (Short-wave Infrared). Each band will have 23.5 m as spatial resolution. Date of the pass of satellite images are given in Table 4.2. False colour composite (FCC) images which prepared from satellite images with available time of pass are shown in Fig. 4.2.

Original Elevation-area-capacity curve and reservoir level on satellite date of pass were acquired from Kallada dam authorities.

Software used for image processing and area of water spread calculation are as ERDAS IMAGINE (Image processing software) and ArcMap (Area calculation).

4.4 Methodology

The main intention involved here is to extract the area of water spread at different elevation in a reservoir, i.e. between full reservoir level (FRL) and minimum draw-down level (MDDL) with the available cloud-free satellite imageries. Flowchart of the methodology adopted in the present study is shown in Fig. 4.3.

Firstly, select the satellite imagery which is free from cloud cover. The data was acquired from National Remote Sensing Centre (NRSC), Hyderabad. The geo-referenced images were utilised to get normalised difference water index (NDWI).

In the visible region (0.4–0.7 μm) of spectrum the significance of transmission, absorption and coefficient of reflection in the water is very less. In the near-IR band, absorption of wavelengths in water increases briskly, whereas both the factor of reflection and transmission is less. With the usage of different signatures of spectral land use types, water pixel extracted from other land use pixels such as soil, vegetation, barren land, etc.

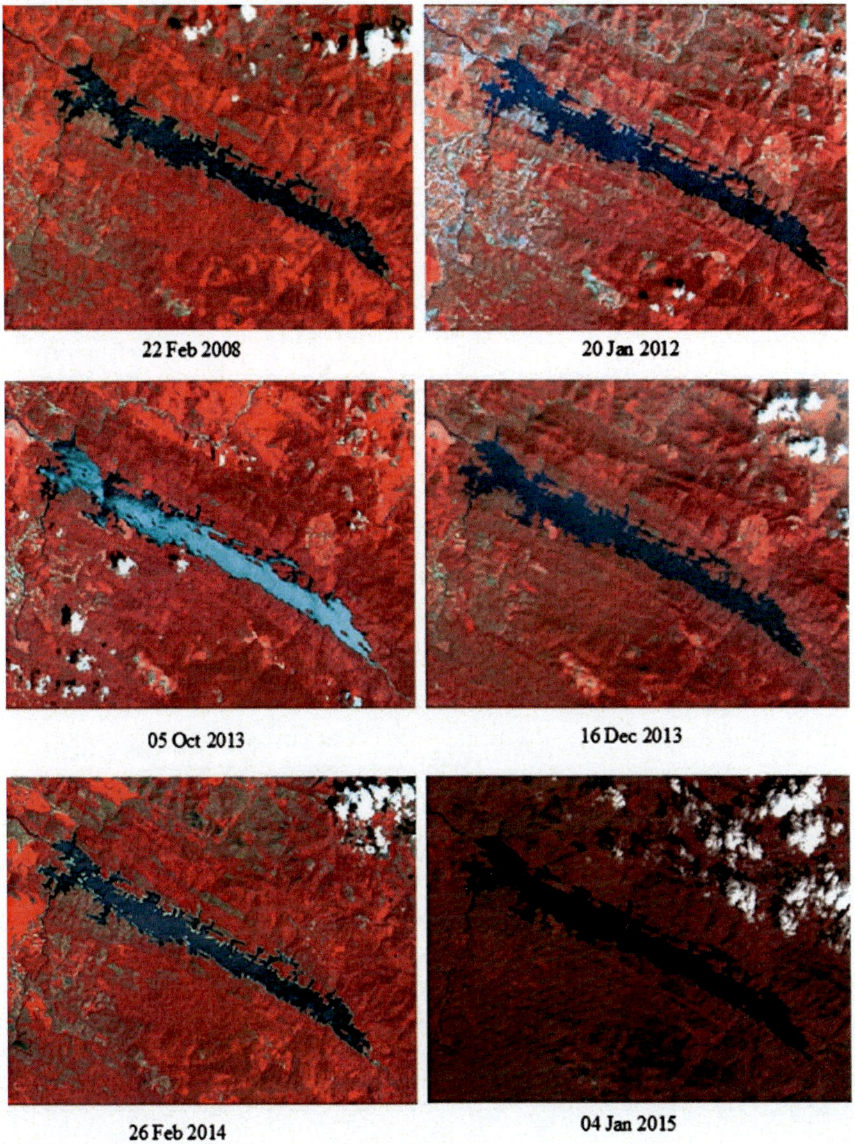

Fig. 4.2 False colour composite of Thenmala reservoir

The normalised difference water index (NDWI) will be accustomed to determine the pixels of water within the imagery. Water absorbs maximum radiation in the NIR region.

The NDWI image processing donc as below:

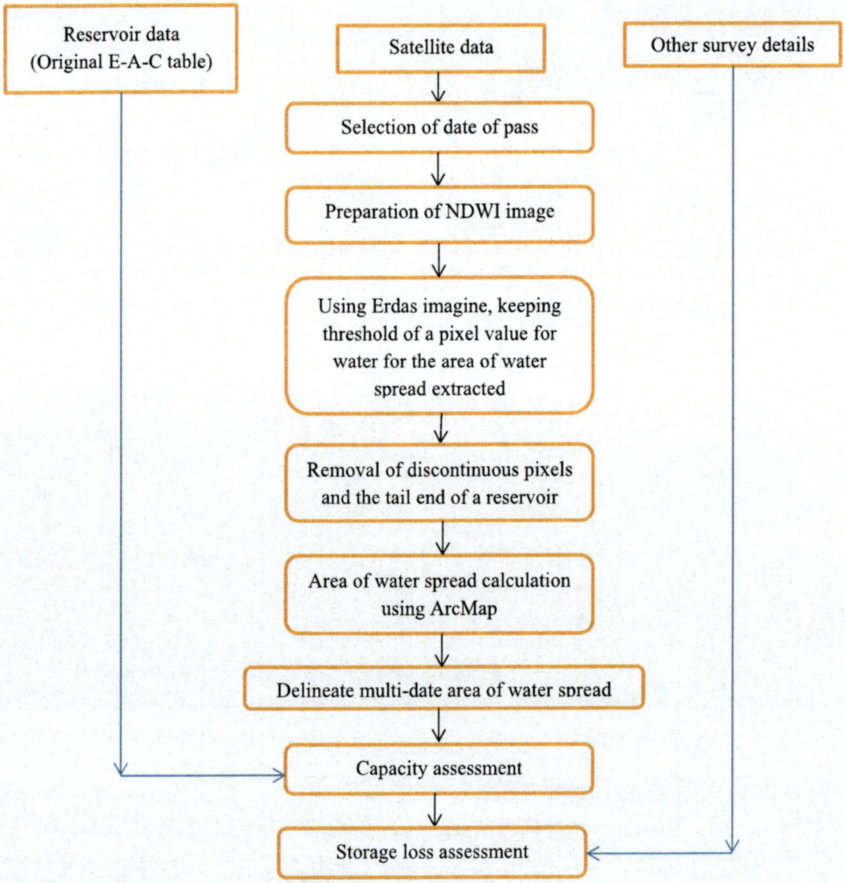

Fig. 4.3 Flowchart of the methodology

$$\text{i.e., NDWI} = \frac{\text{Green} - \text{NIR}}{\text{Green} + \text{NIR}} \quad \text{OR} \quad \frac{\text{BAND2} - \text{BAND4}}{\text{BAND2} + \text{BAND4}}.$$

Selection of the above wavelengths enhances the features of water as mentioned below.

1. Intensifies the distinctive reflectivity in water attributes with the usage of the green band;
2. Curtailment the less reflectivity of Near-IR in water attributes; and
3. Considering the benefit of the high reflectivity of Near-IR by soil and vegetation attributes.

NDWI image can visualise for water pixels and nonwater pixels. So from this NDWI image, all water pixels of the study area are extracted and processed and the area of water spread is estimated. The same process is repeated with the satellite imageries of the different date of pass.

The calculated area unit brought in this operation are used to calculate the volume between 2 consecutive elevations by cone formula. The volume of water (V) between 2 sequential spread $A1$ and $A2$ and elevation difference (H) is calculated as

$$V = H * \left(A1 + A2 + \sqrt{}A1 * A2\right)/3$$

The calculated accumulated capability of the reservoir has obtained by adding calculated volumes between consecutive intervals (Ninija Merina et al. 2016).

4.5 Results and Discussion

Extracted areas of water spread at different date of pass are given in Fig. 4.4 and the calculated area of water spread from satellite imagery is given in Table 4.3.

Extensive care should be taken in the removal of discontinuous water pixels and the water pixels at the downstream of the reservoir, specifying the tail end of the reservoir. There may be a chance of misinterpretation of area estimation if careful editing is not done.

Using the water spread area extracted from imagery, capacity of reservoir is calculated using cone formula and cumulative of calculated capacity is given in Table 4.4.

Live capacity lost due to sedimentation for the survey year 2014 is 44.063 Mm^3 for a span of 30 years, i.e. from 1985 to 2014 between elevation (106.72) and FRL (115.82). Live capacity lost due to sedimentation is given in Table 4.5.

4.6 Conclusions

The accurate estimation of reservoir sedimentation is very much essential to determine the storage capacity of reservoirs and there by effiecient operation of reservoirs. Satellite data is very much useful in this aspect. In this study, Tenmala Dam, Kollam, Kerala, India is taken as the case study to estimate the reservoir sedimentation using satellite data and to estimate the revised elevation capacity curve. In the year 2014, Thenmala dam live storage is calculated as 379.890 Mm^3. In a span 30 years, i.e. from 1985 to 2014 loss of reservoir capacity is 10.39% observed in the zone of live storage. The calculated rate of average siltation was 2.676 mm/year.

The major limitations of remote sensing technique are as follows:

- As the water flucutation in a reservoir can take place from penstock level to full reservoir level. So area of water spread extraction from imagery is restricted only to the zone of live storage, below that is not possible.

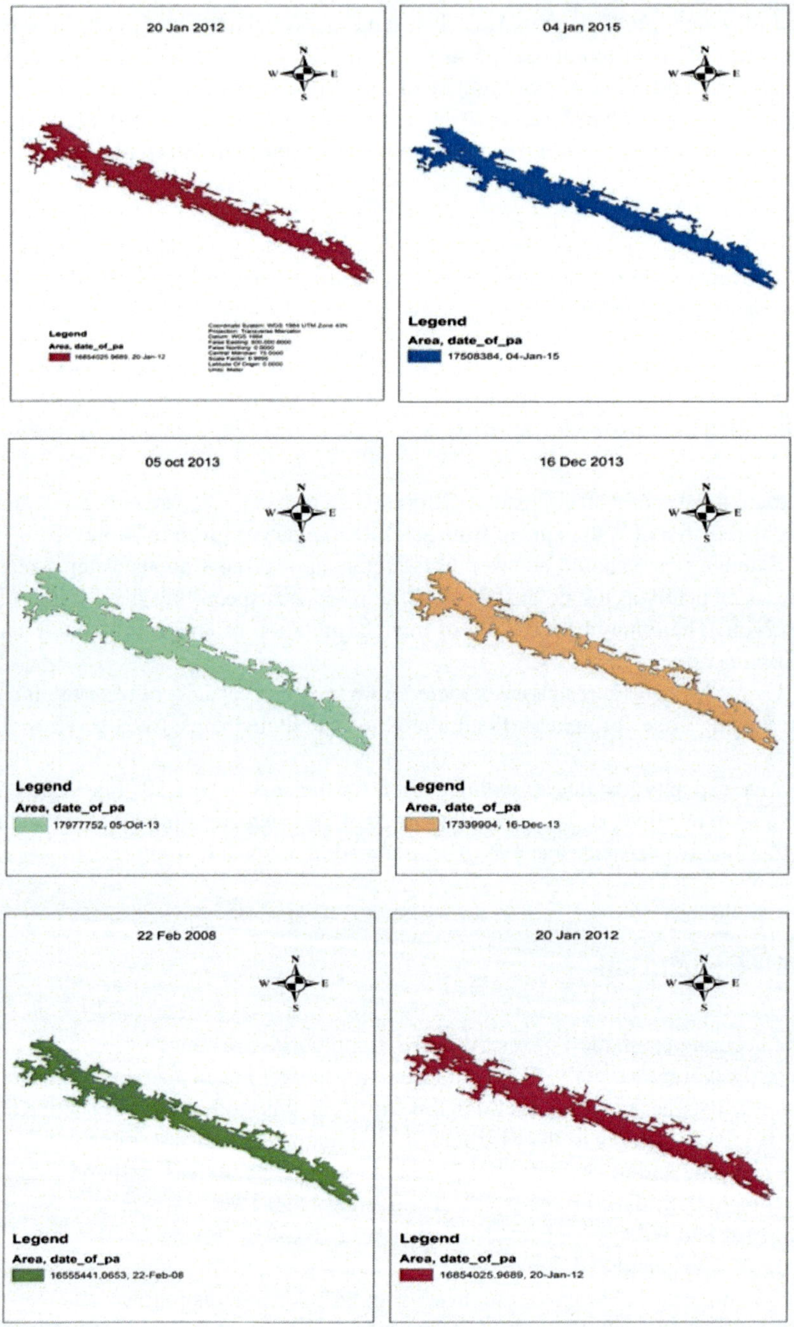

Fig. 4.4 Extracted area of water spread with date of the pass of Thenmala Reservoir

Table 4.3 Area of water spread estimated from satellite data

Date of pass	Elevation (m)	Area (Mm^2)
22 February 2008	109.3	16.555441
20 January 2012	112.22	16.854026
05 October 2013	115.15	17.977752
16 December 2013	114.05	17.339904
26 February 2014	106.72	15.389784
04 January 2015	114.25	17.508384

Table 4.4 Comparsion of calculated capacity with the original capacity

Elevation (m)	Original WSA (Mm^2)	Calculated WSA (Mm^2)	Original volume (Mm^3)	Calculated volume (Mm^3)	original cum. Capacity (Mm^3)	Calculated cum. capacity (Mm^3)
106.72	18.828	15.390			227.142	227.142
109.3	20.325	16.555	50.681	41.159	277.823	268.301
112.22	21.683	16.854	61.337	48.728	339.160	317.028
114.05	22.455	17.340	40.341	31.255	379.501	348.284
114.25	22.516	17.508	4.518	3.481	384.019	351.765
115.15	22.834	17.978	21.149	15.952	405.168	367.717
115.82	23.239	18.397	18.785	12.173	**423.953**	**379.890**

Table 4.5 Live capacity lost due to sedimentation

	Original data (1985)	Survey year (2014) SRS
Live capacity in mm^3 between RL and FRL	423.953	379.890
Sediment deposited in two consecutive surveys in Mm^3		44.063
Period in years		30
Silt deposition between two consecutive surveys Mm^3/year		1.469
Capacity loss		
– Mm^3		44.063
– % to original capacity		10.39

- Lack of accessibility in cloud free imageries throughout the operation of reservoir.

References

Goel MK, Jain SK, Agarwal PK (2002) Assessment of sediment deposition rate in Bargi Reservoir using digital image processing. J Hydrol Sci 47(5):81–92

Ijam AZ, Al-Mahamid MH (2012) Predicting Sedimentation at Mujib Dam Reservoir in Jordan. Jordan J Civ Eng 6(4):448–463

Jain SK, Singh P, Seth SM (2002) Assessment of sedimentation in Bhakra Reservoir in the western Himalayan region using remotely sensed data. Hydrol Sci J 47(2):203–212

Jeyakanthan VS, Sanjeevi S (2013) Capacity survey of Nagarjuna Sagar reservoir, India using linear mixture modal (LMM) approach. Int J Geomat Geosci 4(1):186–194

Narasayya K, Roman UC, Sreekanth S, Jatwa S (2013) Assessment of Reservoir sedimentation using remote sensing satellite imageries. Asian J Geoinform 12(4):172–180

Ninija Merina R, Sashikumar MC, Rizvana N, Adlin R (2016) Sedimentation study in a reservoir using remote sensing technique. Appl Ecol Environ Res 14(4):296–304

Pandey A, Chaube UC, Mishra SK, Kumar D (2016) Assessment of reservoir sedimentation using remote sensing and recommendations for desilting Patratu Reservoir India. Hydrol Sci J 61(4):711–718

Reservoir sedimentation assessments method a Handbook, Bureau of Indian Standards, IS 13665:1993

Yeo I-Y, Lang M, Vermote E (2014) Improved understanding of suspended sediment transport process using multi-temporal landsat data: a case study from the old woman Creek Estuary (Ohio). IEEE J Sel Top Appl Earth Obs Remote Sens 7(2):636–647

Chapter 5
Regionalisation of Watersheds Using Fuzzy C Means Clustering Algorithm in the West Flowing River of Kerala

Thottungal Krishnankutty Drissia, Vinayakam Jothiprakash, and Alayil Bahuleyan Anitha

Abstract Regionalisation of the watersheds, defined as the grouping of statistically similar watersheds is generally carried out for regional flood frequency analysis. This also helps in rainfall-runoff modelling. In this study, the catchment area of 43 stream gauge in the west flowing rivers of Kerala is grouped using fuzzy c means (fcm) clustering algorithm. The attributes in the feature vector and number of clusters are two deciding factors, which has no particular guidelines. Hence, three sets of feature vectors from the catchment characteristics formed for analysis. The first set includes catchment area (A), main channel length (Lc), the circulatory ratio (Rc), average rainfall (P), the total length of streams (Lu), the total number of streams (Nu). Lc, A, Rc, P, Lu, Nu, *L*-CV (L coefficient of variance) forms the second group and Lc, A, Rc, P, Lu, Nu, *L*-CV, latitude and longitude forms the third group. The fuzzy c means clustering was carried out with 4, 5 and 6 cluster numbers for each feature vector. The FCM clustering partitions the rescaled feature vector by minimising the objective function. Clusters were validated using cluster validity indices such as partition coefficient (PC), classification entropy (CE), partition index (SC), separation index (S) and Xie and Beni's index (XB). L moments heterogeneity test (H Test) was used to test the homogeneity of each cluster. It was found that five clusters formed from the second group of feature vector with attributes Lc, A, Rc, P, Lu, Nu, *L*-CV produced homogeneous regions.

Keywords Fuzzy c means cluster · L moment · Heterogeneity test · Validity index · Regionalisation

T. K. Drissia (✉) · V. Jothiprakash · A. B. Anitha
Indian Institute of Technology Bombay, Powai, Maharashtra 400076, India

V. Jothiprakash
e-mail: vprakash@iitb.ac.in

T. K. Drissia · A. B. Anitha
Centre for Water Resources Development and Management, Kozhikode, Kerala 673571, India

R. Jha et al. (eds.), *Water Resources Management and Reservoir Operation*, Water Science and Technology Library 107,
https://doi.org/10.1007/978-3-030-79400-2_5

5.1 Introduction

Regionalisation of watersheds is the first step in regional flood frequency analysis, which identifies regions behaving in a hydrologically similar fashion (Nathan and McMahon 1990). In addition, regionalisation helps in developing the rainfall-runoff model for that particular region. Amongst the many statistical techniques used for regionalisation includes L moment method (Hosking 1990), the region of influence approach (Burn 1990), canonical correlation analysis (Ouarda et al. 2001), ward's cluster method (Jingyi and Hall 2004), hierarchical and trellis method (Chérif and Bargaoui 2013).

Clustering is defined as the classification of data into groups so that the elements in that group process some similar characteristics. Many clustering techniques have been used in regionalisation. Some of the clustering algorithms are the k means algorithm and the Fisher algorithm (Hartigan 1989). From a hydrological point of view, clustering attempts to identify a cluster or group of catchments with similar catchment characteristics and are dissimilar with other clusters (Burn and Goel 2000). Two types of clustering algorithm were defined in the literature. One is partitional clustering and the other hierarchical clustering (Rao and Srinivas 2008). Hierarchical clustering gives nested sequence of partitions, whereas partitional clustering produce partition of the data (Rao and Srinivas 2008). Soft computing techniques employed in regionalisation are self-organising map (SOM), artificial neural network (ANN) (Kumar and Chatterjee 2015), fuzzy c means (fcm) algorithm (Rao and Srinivas 2006; Budayan et al. 2009). A comparison of the artificial neural technique and fuzzy clustering technique shows that fuzzy clustering techniques were better (Kumar and Chatterjee 2015). Two fuzzy clustering techniques are available, Fuzzy c means (fcm) clustering algorithm and subtractive algorithm (Bataineh et al. 2011). In fuzzy clustering of watersheds, one watershed can belong to more than one cluster. In the real case , one watershed may resemble more than one group of catchments. Hence, fuzzy c means (fcm) clustering gives more fruitful results for catchment classification.

Though many works of literature were reported on the application of soft computing techniques in rainfall-runoff modelling, few were reported in regional flood frequency analysis. In this context, the application of the fuzzy c means algorithm (fcm) in the regionalisation of watersheds in west -flowing rivers (WFR) of Kerala was attempted. WFR consists of numerous small and medium rivers that have high variation in topography and climate. Hence, identifying homogeneous regions in this part of the country is challenging.

5.2 Study Area and Data

West flowing rivers (WFR) of Kerala is the study area considered for the analysis. WFR of Kerala forms a major part of the zone 'WFR from Tadri to Kanyakumari', a zone defined by the Central Water Commission (CWC), Govt. of India. This region has 41 WFR that originates from Western Ghats and drains to Arabian Sea. Kerala receives an average annual rainfall of 3000 mm mainly by two monsoon seasons, southwest and northeast.

43 watersheds in WFR were selected based on the catchment area of stream gauge stations maintained by the CWC, (20 stations) and Water Resources Department, Government of Kerala (23 stations) (Table 5.1). The boundary and drainage of the

Table 5.1 Details of watersheds

S. No.	Rivergauging	Catchment area	River basin
1	Manjeswaram	25.44	Manjeswar
2	Anakkallu	166.25	Uppala
3	Shiriya d/s	322.56	Shiriya
4	Shiriya u/s	348.00	Shiriya
5	Madhur	66.04	Mogral
6	Moonnamkadavu	216.80	Chandragiri
7	Erivanjipuzha	957.00	Chandragiri
8	Kakkadavu	276.50	Karingode
9	Mangara	109.60	Kuppam
10	Irude	189.63	Valapattanam
11	Palapuzha	237.25	Valapattanam
12	Perumannu	1070	Valapattanam
13	Kannavam	60.75	Anjarakandy
14	Meruvambai	180.00	Anjarakandy
15	Kuttiyadi	238	Kuttiyadi
16	Kollikkal	34.82	Korapuzha
17	Chaliyar	386.00	Chaliyar
18	Kanjirapuzha	64.00	Chaliyar
19	Kuniyil	1876	Chaliyar
20	Karathodu	750	Kadalundi
21	Ambarampalayam	950	Bharathapuzha
22	Kumbidi	5755	Bharathapuzha
23	Mankara	2,775	Bharathapuzha
24	Pudur	1313	Bharathapuzha

(continued)

Table 5.1 (continued)

S. No.	Rivergauging	Catchment area	River basin
25	Pulamanthole	940	Bharathapuzha
26	Arangaly	1342	Chalakudy
27	Karuvannur	725.00	Karuvannur
28	Ambalakadavu	1160.00	Chalakudy
29	Neeleeswaram	4234	Periyar
30	Kalampur	405	Moovattupuzha
31	Ramamangalam	1208	Moovattupuzha
32	Kidangoor	615	Meenachil
33	Teekoy	57.00	Meenachil
34	Manimala	490.00	Manimala
35	Kallooppara	731	Manimala
36	Malakkara	1713	Pamba
37	Thumpamon	810	Achankoil
38	Kalleli	419.00	Achenkoil
39	Kollakkadavu	952.71	Achenkoil
40	Punalur	870.00	Kallada
41	Pattazhi	1210	Kallada
42	Ayilam	540	Vamanapuram
43	Ottasekharamangalam	247.35	Neyyar

watersheds were digitised. This was used for estimating the catchment characteristics. The details of watersheds are given in Fig. 5.1. The weighted average rainfall of the watersheds was also estimated.

5.3 Methodology

Clustering of watersheds was carried out using fcm clustering algorithm. The methods are explained in following sections.

5.3.1 Fuzzy Clustering Algorithm

In fuzzy c means clustering the attributes are selected to make a matrix of gauging stations and attributes (Rao and Srinivas 2008). Based on the correlation between the topographic parameters and peak discharge eleven attributes were selected for creating feature vector. The parameters are the area of catchment up to stream gauge

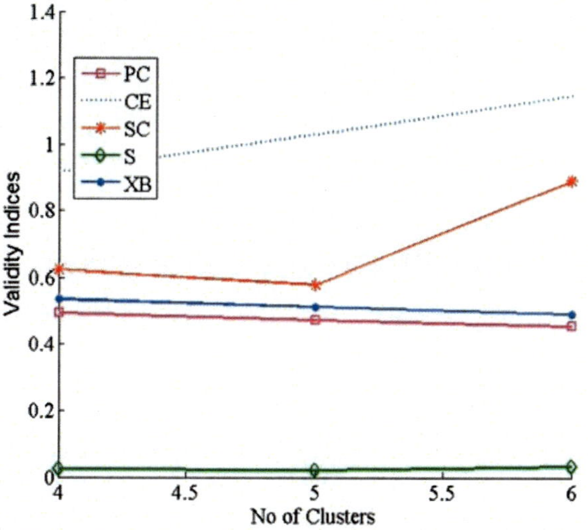

Fig. 5.1 Value of validity indices for the clusters formed out of feature vector Lc, A, Rc, P, Lu, Nu

station, main channel length, maximum basin length, shape factor, the circulatory ratio, relief ratio, the total length of streams, average annual rainfall, the total number of streams, latitude and longitude. The details of topographic parameters are given in Sect. 5.4.1. The coordinates included in the feature vector accounts the proximity of the stations. The feature vector is rescaled for bringing the values in 0–1 scale. For rescaling the values of the parameters, Eq. 5.1 is used. Rescaling is necessary, as there is difference in magnitude, unit and importance.

$$x = \frac{x - x_{(\min)}}{x_{(\max)} - x_{(\min)}} \tag{5.1}$$

This feature vector is used for clustering using fuzzy c means algorithm. It is given by $Y_k = [y_{1k} + y_{2k} + \cdots + y_{nk}]$ where y_{ik} is denotes the value of attribute i in y_k.

The rescaled feature vectors represented as a matrix X provided in Eq. 5.2.

$$X = \begin{bmatrix} x_{11} \ldots & x_{1N} \\ \ldots & \ldots \\ x_{n1} & x_{nN} \end{bmatrix} \tag{5.2}$$

The centroids of the clusters (c) are represented by $V = (v_1, \ldots, v_c)$ in which v_1 is the centroid of cluster 1. The following objective function is minimised to partition the data X to c clusters.

$$J(UV : X) = \sum_{i=1}^{c} \sum_{k=1}^{N} (u_{ik})^m d^2(x_k, v_i) \tag{5.3}$$

Subjected to following constraints

$$\sum_{i=1}^{c} U_{ik} = 1 \quad \forall k \in (\{1, \ldots, N\}) \tag{5.4}$$

$$0 < \sum_{k=1}^{N} U_{ik} < N \quad \forall i \in \{1, \ldots, c\} \tag{5.5}$$

where $u_{ik} \in [0, 1]$–membership of the kth rescaled feature vector x_k in the ith fuzzy cluster.

U is the fuzzy partition matrix which contains the membership of each rescaled feature vector in each fuzzy cluster.

The parameter $\mu \in [1, \infty]$ refers to the weight exponent for each fuzzy membership.

$$U = \begin{bmatrix} u_{11} \ldots u_{1k} \ldots u_{1N} \\ . \quad\quad . \quad\quad . \\ u_{i1} \quad u_{ik} \quad u_{iN} \\ . \quad\quad . \quad\quad . \\ u_{c1} \quad u_{ck} \quad u_{cN} \end{bmatrix} \tag{5.6}$$

The distance from kth rescaled feature vector x^k to the centroid of ith cluster v_i

$$d^2(x_k,\ v_i) = (x_k - v_i)^T A_i (x_k - v_i) \tag{5.7}$$

where the norm A_i is a positive definite symmetric matrix associated with cluster i. For estimation of Euclidean distance between x_k and v_i, $A_i = I\ \forall i$, where I is a unit matrix.

5.3.2 Validity Index

Many cluster validity index are available in literature. Some of the indices relevant to fuzzy c means clustering are applied in this study. This include partition coefficient (PC), classification entropy (CE), partition index (SC), separation index (S) and Xie and Beni's index (XB).

The partition coefficient formulated by Bezdek (1974) uses only membership values and is given by Eq. 5.8.

$$V_{\mathrm{PC}}(U) = \frac{1}{N}\sum_{i=1}^{c}\sum_{k=1}^{N}(u_{ik})^2 \tag{5.8}$$

Partition entropy or classification entropy (CE) (Bezdek et al. 1984)

$$V_{\mathrm{CE}}(U) = -\frac{1}{N}\left[\sum_{i=1}^{c}\sum_{k=1}^{N} u_{ik}\log_a(u_{ik})\right] \quad (5.9)$$

Partition index (SC) (Sote and Pande 2014)

$$\mathrm{SC} = \sum_{i=1}^{c} \frac{\sum_{j=1}^{N}\left(u_{ij}\right)^m x_j - v_i^2}{N_i \sum_{k=1}^{c} v_k - v_i^2} \quad (5.10)$$

Separation Index (S)

$$S = \frac{\sum_{i=1}^{c}\sum_{j=1}^{N}\left(u_{ij}\right)^m x_j - v_i^2}{N \min_{ij} x_j - v_i^2} \quad (5.11)$$

Xie and Beni's index (XB) (1991)

$$\mathrm{XB}(c) = \frac{\sum_{k=1}^{c}\sum_{i=1}^{n} u_{ik}^2 x_i - v_k^2}{nx \min_{i\neq j}\left(v_i - v_j\right)} \quad (5.12)$$

5.3.3 L *Moments Heterogeneity* H *Test*

Once the clusters are formed, the homogeneity in each cluster is checked with heterogeneity test (H test) formulated by Hosking and Wallis (1997). Before H test, discordancy test was conducted to identify any discordant sites. For performing both the tests L moment ratio such as L-covariance (L-CV), L skewness (L-skew) and L kurtosis (L-kurt) were estimated based on the equations given in Hosking and Wallis (1997). The heterogeneity test (H test) estimated is as follows.

$$H = \frac{(V - \mu_{v)}}{\sigma_v} \quad (5.13)$$

where V is the weighted standard deviation of L-CV of all the stations.

μ_v and σ_v are mean and standard deviation of simulated weighted standard deviation of L-CV.

For this, 500 realisations were simulated using four- parameter kappa distribution. The four parameters representing mean (χ), variance (α) and shape (k, h) were estimated by fitting kappa distribution to regional average L moment ratios 1, t^R, $t3^R$, $t4^R$. The regions with $H <$ one regarded as "acceptability homogeneous", H value in the range 1–2 as "possibly homogeneous" and $H >$ two "definitely heterogeneous".

5.4 Results and Discussion

Initial screening was carried out by discordancy test and was found that two stations were discordant. The stations are Manjeswaram and Madhur which are removed before regionalisation.

5.4.1 *Parameter Selection for Feature Vector*

The catchment characteristics are correlated with peak discharge of gauging station to find out the parameters that influence the peak discharge. The linear aspects of catchment characteristics include mainstream length (Lc), maximum basin length (Lb), length of overland flow (Lo), bifurcation ratio (Rb), stream length ratio (Lsr), total length of streams (L) and number of streams (Nu). The areal aspects as area of catchment upto gauging station (A), form factor (Ff), shape factor (Fs), circulatory ratio (Cr), elongation ratio (Re), drainage density (Dd), stream frequency (Sf), constant of channel maintenance (Cc). Apart from the above parameters, relief ratio (Rr) weighted average rainfall (P) and *L* moment variance (*L*-CV) are also considered. All the parameters except rainfall and *L*-CV are correlated with peak discharge. The parameters with high-correlation coefficient as given in Table 5.2 taken for creating feature vector.

Table 5.2 Regression coefficient

Catchment characteristics	R
Maximum basin length (Lb, km)	0.621
Main stream length (Lc, km)	0.637
Length of overland flow (Lo, km)	0.131
Mean bifurcation Ratio (Rb)	0.092
Mean stream length ratio (Lsr)	0.048
Total stream length, km (*Lu*)	0.773
Number of streams (count) (Nu)	0.696
Catchment area (A, km^2)	0.711
Form factor (Ff)	0.098
Shape factor (Fs)	0.391
Circularity ratio (Rc)	0.337
Elongation ratio (Re)	0.092
Drainage density (Dd, km/km^2)	0.093
Stream frequency (Sf)	0.007
Constant of channel maintenance (Cc)	0.131
Relief ratio (Rr)	0.194

The parameters with high-regression coefficient as in Table 5.1 are catchment area (A), maximum basin length (Lb), mainstream length (Lc), total stream length (Lu), number of streams (Nu), shape factor (Fs) and circulatory ratio (Rc). The coordinates (lat and long) are also considered to understand the influence of proximity on cluster formation.

5.4.2 Fuzzy C Means Clusters

The input required for fcm clustering are attributes for feature vector and number of clusters. There are no guidelines for the selection of attributes and the number of clusters. Initially, all the parameters were used as the feature vector, but no homogenous regions formed. Thereafter, the feature vector was created out of the rescaled value of the ten parameter based on the regression coefficient, A, Lb, Lc, L, Nu, Fs, Rc, lat, long, P (Table 5.2). The maximum basin length (Lb) and main stream length (Lc) are similar characteristics and hence, only Lc is selected for the clustering. Likewise, only total stream length (Lu) was selected between Lu and Nu. Shape factor (Fs) and circulatory ratio (Rc) are related to perimeter and catchment area and hence only one is selected. Based on these three sets of feature vectors are [Lc, A, Rc, P, Lu, Nu], [Lc, A, Rc, P, Lu, Nu, *L*-CV] and [Lc, A, Rc, P, Lu, Nu, L-CV, lat, long]. Another input required is optimum number of clusters, which is explained in following section.

5.4.3 Validity Index

Validity indices such as partition coefficient (PC), classification entropy (CE), partition index (SC), separation index (S) and Xie and Beni's index (XB) were estimated for three cluster centres 4, 5 and 6. Partition coefficient measures the quantity of overlap between clusters. The high value of PC and low value of CE indicate minimum overlap between the clusters. The values of validity index estimated for the three set of feature vectors are given Figs. 5.1, 5.2 and 5.3. But the values estimated from the clusters says that there are some overlapping amongst the clusters.

5.4.4 L Moments H Test

The *H* test was estimated for three sets of feature vectors with various combinations of the number of clusters. The *H* value of each cluster is given in Table 5.3. In the first set of the feature vector [Lc, A, Rc, P, Lu, Nu], the *H* value was mostly above 2 indicating a heterogeneous region. The third feature vector, [Lc, A, Rc, P, Lu, Nu, L-CV, lat, long] also produced heterogeneous regions. However, the *H* value

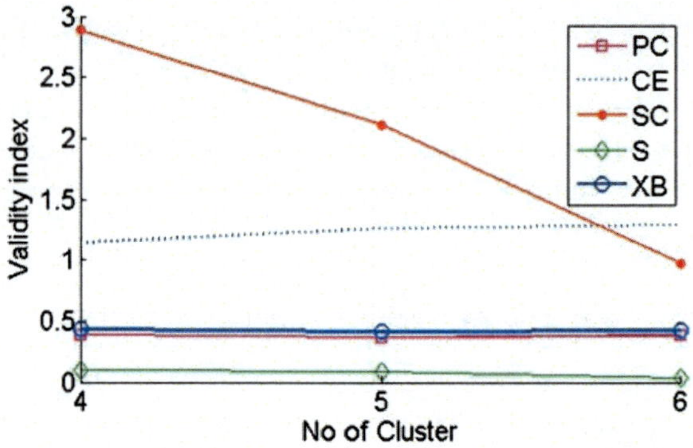

Fig. 5.2 Value of validity indices for the clusters formed out of feature vector [Lc, A, Rc, P, Lu, Nu, *L*-CV]

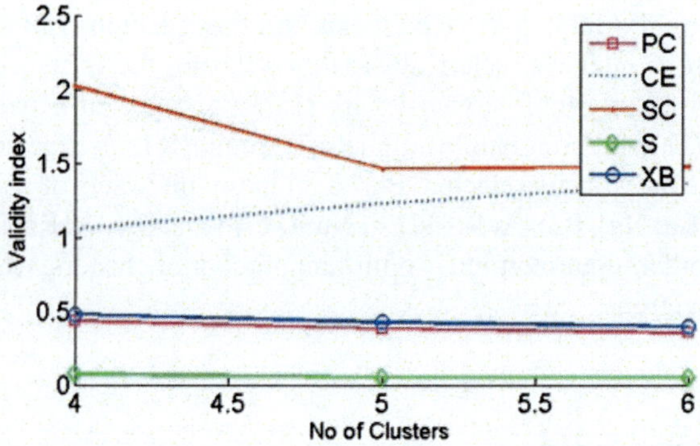

Fig. 5.3 Value of validity indices for the clusters formed out of feature vector [Lc, A, Rc, P, Lu, Nu, lat, long]

of second feature vectorwith five number of clustersis either below one or between one and two indicating"acceptability homogeneous" and "possibly homogeneous". The inclusion of the *L* Coefficient of variance has a good impact on the formation of clusters of watersheds. Interestingly, the inclusion of coordinates does not provideho-mogeneous clusters. It can be stated that in WFR nearness of the watersheds is not an influencing factor in creating homogeneous regions. Five homogeneous clusters were formed with Lc, A, Rc, P, Lu, Nu, *L*-CV as attributesin the feature vector.

The location of the watersheds in each region (zone) is given in Fig. 5.4. However, there are few nearby watersheds that form the cluster, there is no continuity in the clustering. In addition, the watersheds in each cluster are distributed all over the study area. Defuzzification was not carried out as it was stated by (Rao and Srinivas 2008) that considerable effort should be taken for adjusting hard clusters for forming homogeneous regions. It was suggested to take the final outcome of the fcm clustering algorithm as fuzzy clusters.

Table 5.3 H value for each cluster and feature vector

S. No	Feature vector	Cluster name	Number of cluster		
			4	5	6
1	Lc, A Rc, P, Lu, Nu	C1	3.29	2.51	2.7
		C2	NC	3.76	NC
		C3	2.93	2.89	3.55
		C4	NC	**−0.83**	NC
		C5		NC	**0.88**
		C6			2.68
2	Lc, A, Rc, P, Lu, Nu, *L*-CV	C1	NC	**1.79**	2.13
		C2	2.24	**0.78**	**1.82**
		C3	**0.53**	**0.59**	2.58
		C4	2.2	**1.36**	NC
		C5		**1.54**	**0.9**
		C6			NC
3	Lc, A, Rc, P, Lu, Nu, *L*-CV, lat, long	C1	NC	2.68	2.19
		C2	2.09	2.16	NC
		C3	2.39	4.76	**1.52**
		C4	**1.38**	**1.59**	2.60
		C5		**1.57**	**1.56**
		C6			2.13

NC-kappa parameter not converged

5.5 Conclusion

Regionalisation of watersheds was carried out using the fuzzy c means algorithm. The major inputs needed for fcm clustering are the number of clusters and parameters required for the feature vector. Out of 20 catchment characteristics selected, seven parameters have a good correlation with peak discharge. Combinations of the parameters are attempted and found that feature vector with catchment area (A), mainstream length (Lc), circulatory ratio (Rc), weighted average rainfall (P), total stream length (Lu), the total number of streams (Nu) and *L* moment variance (*L*-CV) produce five homogeneous regions. The regions are not continuous though few watersheds are nearby in each cluster. The selection of attributes for feature vector has not followed any particular guidelines, hence, further analysis on procedures for selection of suitable attributes may be attempted in future studies. Flood frequency relationships can be developed for each cluster using the index flood method. This will help in estimating flood frequency for the ungauged sites.

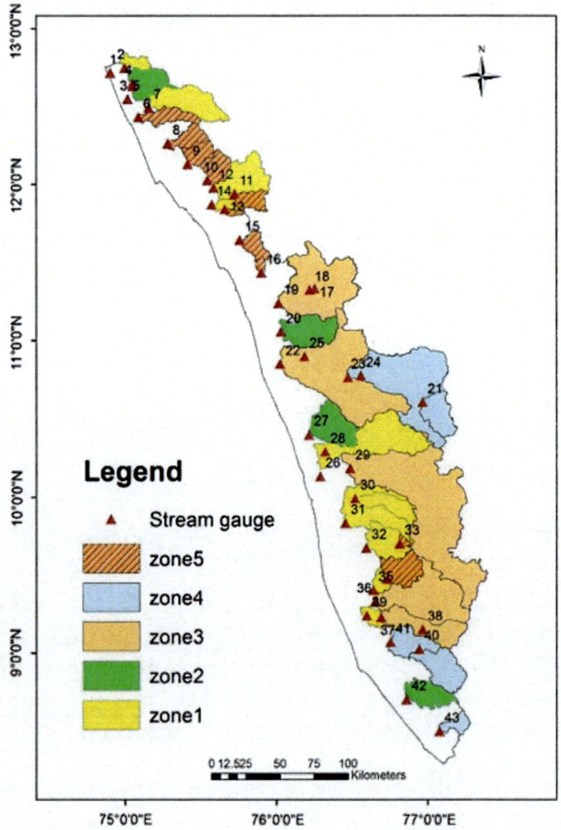

Fig. 5.4 Homogeneous Regions formed using fuzzy c means clustering algorithm

Acknowledgements Authors are thankful to the Central Water Commission, Government of India and Water Resources Department, Government of Kerala for sharing the discharge data.

References

Bataineh KM, Naji M, Saqer M (2011) A comparison study between various fuzzy clustering algorithms. Jordan J Mech Indust Eng 5(4):335–343

Bensaid AM, Hall LO, Bezdek JC, Clarke LP, Silbiger ML, Arrington JA, Murtagh RF (1996) Validity-guided (Re)Clustering with applications to image segmentation. IEEE Trans Fuzzy Syst 4:112–123

Bezdek JC, Ehrlich R, Full W (1984) FCM: The fuzzy c-means clustering algorithm. Comput Geosci 10(2–3):191–203

Budayan C, Dikmen I, Birgonul MT (2009) Comparing the performance of traditional cluster analysis, self-organizing maps and fuzzy C-means method for strategic grouping. Expert Syst Appl 36(9):11772–11781. https://doi.org/10.1016/j.eswa.2009.04.022

Burn DH (1990) Evaluation of regional flood frequency analysis with a region of influence approach. Water Resour Res 26(10):2257. https://doi.org/10.1029/WR026i010p02257

Burn DH, Goel NK (2000) The formation of groups for regional flood frequency analysis. Hydrol Sci J 45(February):97–112. https://doi.org/10.1080/02626660009492308

Chérif R, Bargaoui Z (2013) Regionalisation of maximum annual runoff using hierarchical and Trellis methods with topographic information. Water Resour Manage 27:2947–2963. https://doi.org/10.1007/s11269-013-0325-0

Hartigan JA (1989) Clustering algorithms. Wiley, pp 331–335. https://doi.org/10.1002/0471725382

Hosking JRM (1990) L moments: analysis and estimation of distributions using linear combination of order statistics. J Roy Stat Soc 52(1):105–124

Hosking JRM, Wallis JR (1997) Regional frequency analysis: an approach based on L-moments. Cambridge University Press, New York

Jingyi Z, Hall MJ (2004) Regional Flood Frequency Analysis for the Gan-Ming River Basin in China 296:98–117. https://doi.org/10.1016/j.jhydrol.2004.03.018

Kumar R, Chatterjee C (2015) Regional flood frequency analysis using soft computing techniques. Water Resour Manage, pp 1–7. https://doi.org/10.1007/s11269-015-0922-1

Nathan RJ, McMahon TA (1990) Identification of homogeneous regions for the purposes of regionalisation. J Hydrol 121:217–238. https://doi.org/10.1016/0022-1694(90)90233-N

Ouarda TBMJ, Girard C, Cavadias GS, Bobée B (2001) Regional flood frequency estimation with canonical correlation analysis. J Hydrol 254(1–4):157–173. https://doi.org/10.1016/S0022-1694(01)00488-7

Rao AR, Srinivas VV (2006) Regionalization of Watersheds by Fuzzy Cluster Analysis 318:57–79. https://doi.org/10.1016/j.jhydrol.2005.06.004

Rao AR, Srinivas VV (2008) Regionalization of watersheds. Springer, Berlin

Sote AM, Pande SR (2014) Web based fuzzy clustering analysis. Res Inventy: Int J Eng Sci 4(11):51–57

Xie X, Cui Y (2011) Development and test of SWAT for modeling hydrological processes in irrigation districts with paddy rice. J Hydrol 396(1–2):61–71. https://doi.org/10.1016/j.jhydrol.2010.10.032

Chapter 6
Analysis of Relationship Between Landslides and Rainfall in Karwar, Uttara Kannada District, Karnataka, India

G. Thejashree, K. N. Lokesh, and G. S. Dwarakish

Abstract Landslide is a hazardous geological phenomenon and a common occurrence in hilly regions. It causes loss of life and destruction of property depending on the intensity of its occurrence. Landslide is the movement of mass of rock, earth or debris down the slope. One of the important triggering factors for landslide occurrence is rainfall and is usually considered for slope failure predictions, yet the relationship between rainfall and landslide occurrences is very complex. An attempt is made to understand the relationship between rainfall and landslide in the study area. A devastating landslide occurred in October 2009 in Kadwad village of Karwar taluk in Uttara Kannada district, killing 19 people and burial of several houses. On the same ill fated day, there occurred landslides at 21 more locations in Karwar. The landslide incidence was triggered due to heavy rainfall on 2nd and 3rd October 2009 which resulted in mass movement of debris. For this study, daily precipitation data from Karnataka State Natural Disaster Monitoring Centre/District Disaster Professional, Karwar and landslide records from media and other sources of that particular year were considered for the statistical analysis to understand landslide mechanism. Analysis shows a strong correlation between landslide occurrence and rainfall pattern. The landslide incidence is determined by the magnitude of the short episode of rainfall intensity with the threshold value of above 260 mm in that particular year. Thus the rainfall with less than the minimum threshold value did not cause triggering of landslide whereas landslide occurred above the threshold value. The possible cause for the triggering of Kadwad landslide is not due to saturation of soil caused by antecedent precipitation but it is due to short term intensive rainfall.

Keywords Rainfall · Kadwad landslide · Rainfall pattern · Threshold · Statistical analysis

---

G. Thejashree (✉) · K. N. Lokesh
Department of Civil Engineering, National Institute of Technology Karnataka (NITK), Surathkal, Srinivasnagar Post, Karnataka 575025, India

G. S. Dwarakish
Department of Applied Mechanics & Hydraulics, National Institute of Technology Karnataka (NITK), Surathkal, Srinivasnagar Post, Karnataka 575025, India

R. Jha et al. (eds.), *Water Resources Management and Reservoir Operation*, Water Science and Technology Library 107,
https://doi.org/10.1007/978-3-030-79400-2_6

6.1 Introduction

Landslide is one of the disastrous geological phenomena posing grave threat to life and property. Rainfall is one of the significant factors which causes landslide occurrence especially in regions with extreme seasonal precipitation. Other factors such as climatic factors, geology, geomorphology of the area and human intervention also play a major part in inducing landslides. Much previous research has been carried out by researchers to correlate rainfall and landslide occurrence in hilly terrains. The most important and difficult task for landslide research is to infer how rainfall-induced landslide process works. On the basis of previous studies on information of rainfall data and landslides incidences, the threshold required to provoke landslide initiation can either be related to the significant increasing amount of rainfall continuously or to the intensity of rainfall. The most commonly used method globally for such threshold was developed by Caine (1980) taking into consideration the intensity and duration of rainfall event. Earlier, threshold studies have been carried out by many researchers to establish empirical relation between landslide incidences and total amount of rainfall occurred (Brand et al. 1984; Claessens et al. 2007 and Bai et al. 2014).

Various models have been suggested by several researchers for thresholds of landslide initiation due to rainfall, such as statistical analysis by correlating relationship among intensity of rainfall (I), rainfall duration (D) and cumulative rainfall (E), Logistic regression model, hydrological modeling and quantitative antecedent soil water status (ASWS) models as shown by Caine (1980), Terlien (1998), Lan et al. (2003), Tsai (2008), Sengupta et al. (2010), Rosi et al. (2012) and Bai et al. (2014). The rainfall threshold is calculated by correlating the rainfall amount necessary for landslide initiation with highest intensity of rainfall (Govi et al. 1985). Glade et al. (2000) considered the effect of antecedent rainfall in rainfall threshold for initiation of landslide. The rainfall threshold calculation can be formulated to adapt it in mitigating the consequences of landslide (Jakob and Weatherly 2003). Sengupta et al. (2010) calculated rainfall threshold based on cumulative rainfall and rainfall duration, intensity of rainfall and Mean Annual Precipitation (MAP). In yet another case study analysis on the relationship of landslide occurrence and rainfall was made using antecedent rainfall saturation and regression model. The resulting threshold values can be helpful to those institutions and organizations which deal with failures of slope by evaluating the possibility of occurrence of landslide for diverse rainfall events (Bai et al. 2014). Peruccacci et al. (2017) also attempted to quantify rainfall thresholds for the landslide initiation using cumulative rainfall amount and duration. Landslide triggering phenomena such as rainfall, groundwater conditions like permeability, antecedent degree of saturation and soil characteristics are not well understood. Moreover, prediction of landslides due to rainfall in various regions is difficult owing to lack of availability of requisite data and continuous monitoring of small scale landslides events.

The present study aims to examine the relationship between rainfall event and landslide incidents by statistical analysis of rainfall threshold in landslide initiation.

The analysis was carried out on Kadwad landslide which occurred on October 2nd& 3rd 2009 in Karwar taluk of Uttara Kannada district, Karnataka. The current work is based on the rainfall data and landslide inventory available for Kadwad landslide event in Karwar.

6.2 Study Area

Karwar taluk in Uttara Kannada district of Karnataka forms a part of Western Ghats where the range dips into the sea with peaks emerging as picturesque islands. The geographical area is situated at 14° 50′ 2.12″ N and 74° 10′ 44.50″ E; 14° 50′ 34.25″ N and 74° 10′ 35.24″ E which lies in Survey of India (SOI) Toposheet no. 48 J/1 (Fig. 6.1). The area receives average annual precipitation of approximately 3000 mm. On October 2nd, 2009 between 4:15 PM and 4:30 PM a massive landslide occurred in Zariwad locality of Kadwad village (Fig. 6.3), east and south of Karwar town and Kali River estuary respectively. The aggressive landslide that took place killed 19 people and devastated 9 houses. Also, high intensity precipitation of two consecutive days resulted in subsidiary landslides in 21 locations all along the hills obstructing traffic movement on NH-17. The massive slide in Kadwad village was characterized as debris slide. At some locations, it consisted of debris flow, weathered materials and lateritic boulders while at some places rock fall was also witnessed (Fig. 6.2).

As reported by Geological Survey of India, the landslide of Kadwad is considered as deep seated of more than 20 m depth of failure, large rotational landslide which had a fast rate of movement and having slope angle of approximately 30°.

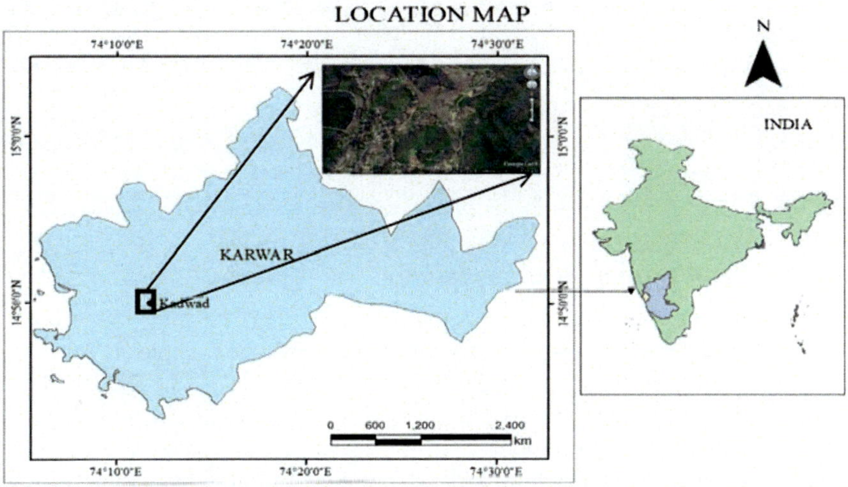

Fig. 6.1 Location of the study area

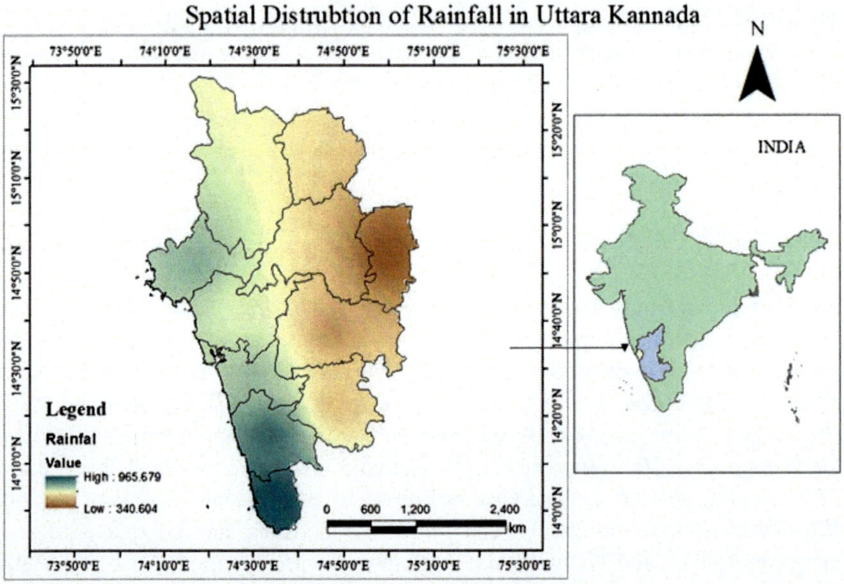

Fig. 6.2 Spatial distribution of rainfall in Uttara Kannada District

Fig. 6.3 **a** & **b** Landslide photographs in Kadwad Village

6.2.1 *Geology and Geomorphology*

Regionally, the area comprises rocks of Peninsular Gneissic Complex (PGC), older granitoids and rocks belonging to the Chitradurga Group under the DharwarSuper-group of Archaean age as well as dolerites of Palaeoproterozoic age intruding older rocks. The rocks have undergone intensive chemical weathering resulting in the formation of laterites, a common rock type in humid tropical environment. In the study area, laterites form a thick mantle capped over underlying granite. Along with granite, the area also possesses saprolite at places and lateritic soil with lithomargic clay is also present. Geomorphologically, the area is low-lying valley region affected by percolation of groundwater. The area is recognized as paleo river channel that flowed into the Kali River in the past which is in north of Zariwad of Kadwad village. This old river channel is still linked to the Kali River and the level of groundwater percolation in the channel fluctuates at regular intervals in respect to high and low tides of the river. Konkan Railway track goes through this paleo-river valley connects between Karwar town with Goa and other places. A number of lineaments are present in the area with high density dendritic drainage pattern where higher stream orders are structurally controlled (Hegde et al. 2014).

6.2.2 *Climate and Rainfall*

The area lies on a coastal strip known as the Monsoon Coast. It is surrounded by the Western Ghats and Arabian Sea on east and west respectively. The area possesses tropical climate with hot summers from March to May where temperature can reach as high as 37 °C. Winter starts from December to February and has mild temperatures of 24–32 °C. The monsoons are experienced in between June and October and receive average annual precipitation of over 3000 mm. Most of the rainfall occurs during the peak monsoon period from June to August .

The rainfall in the area includes both Southeast and Northeast monsoons. During peak monsoon period, rainfall is received from southeast monsoon. Retreating monsoon in the area accounts from October to December and has an average annual rainfall of around 250 mm. In the year 2009, the amount of rainfall in retreating monsoon was more when compared to other years. The spatial distribution of rainfall for Uttara Kannada district is shown in Fig. 6.2.

6.3 Materials and Methodology

Rainfall and landslides are closely associated processes since rainfall is one of the chief driving factors for landslide occurrences in Western Ghats. For present studies, daily precipitation data was obtained from Karnataka State Natural Disaster Monitoring Centre/District Disaster Professional, Karwar for over ten years from 2001 to 2010 was used for statistical analysis. Reported landslide of October 2009 from media and other sources corresponding to rainfall data of that year was analyzed and calculated for rainfall threshold. Threshold is a level either maximum or minimum at which a process tends to experience change. Two thresholdstypes have been defined in the literature:

i. Minimum threshold: it is the minimum stage below which a process does not experience change

ii. Maximum threshold: it is the stage over which a process always experiences change.

To study the interrelationship between rainfall and landslide occurrences threshold can be derived from the amount of rainfall where if it is below the threshold landslide does not occur and if it's above the threshold landslide always occurs with respect to other variable conditions of the area. Empirical models are used to define rainfall threshold as suggested by many researchers (Crosta and Frattini 2001; Aleotti 2004; Guzzetti et al. 2007). Threshold values for rainfall triggered landslides are marked based on the rainfall measurements such as rainfall intensity (I), duration (D) and rainfall event (E) as seen in many literatures. The ratio of the total rainfall received for a given duration and the corresponding duration of time is defined as the Rainfall Intensity and expressed in depth per unit time e.g. mm/hr or cm/hr. Caine (1980) proposed intensity-duration threshold for landslide by taking 73 records globally and derived an empirical form.

$$I = 14.82\ D^{-0.39} \tag{6.1}$$

In addition to I-D relationships, thresholds based on rainfall event (*E*) and Duration (*D*) have also been proposed by (Cannon and Ellen 1985) to predict rainfall induced landslides. As proposed by Caine (1980) a global *E-D* threshold for shallow landslides as:

$$E = 14.82\ D^{0.61} \tag{6.2}$$

Antecedent rainfall measurements along with duration have also been combined to derive thresholds for landslide initiation by many authors (Kim et al. 1991; Terlien 1998; De Vita 2000; Heyerdahl et al. 2003; Gabet et al. 2004; Cardinali et al. 2006). A derivative index called E_{MAP} was proposed considering the cumulative rainfall for the event normalized by the Mean Annual Precipitation (MAP) (Sengupta et al. 2010) for local extent of area.

$$E_{\text{MAP}} = \frac{E}{\text{MAP}} \tag{6.3}$$

6.4 Results and Discussion

Multiple landslides were reported for this area during 2nd and 3rd October 2009. The local rainfall data for the years 2001–2010 was considered for this analysis. Coastal talukas of the district receive comparatively higher rainfall (Fig. 6.2). The mean annual precipitation for the area is around 3277 mm, almost 85% of which is obtained during the southwest monsoon period (June to September), 5–6% during the premonsoon period (January to May) and 8–9% during the northeast monsoon season (October to December). The rainfall records for the study area are available for 24 h intervals. The rainfall variation over the years 2001–2010 is shown in Fig. 6.4a which clearly shows the concentration of rainfall over the monsoon season every year.

A detailed plot for rainfall (mm) for the year 2009 is also presented in Fig. 6.4b with major landslide events marked with red asterisks. Records indicate several occurrences of high daily rainfall but the concentration of landslide occurrences during October indicates a role of antecedent rainfall also. While there are at least 18 days where the daily rainfall exceeded 50 mm, only two of these resulted in landslides. There are number of reported landslides during 2nd to 3rd October 2009 of which major slides were reported in Kadwad village locality. While the rainfall on 3rd October was the peak rainfall for the year and can explain the landslides

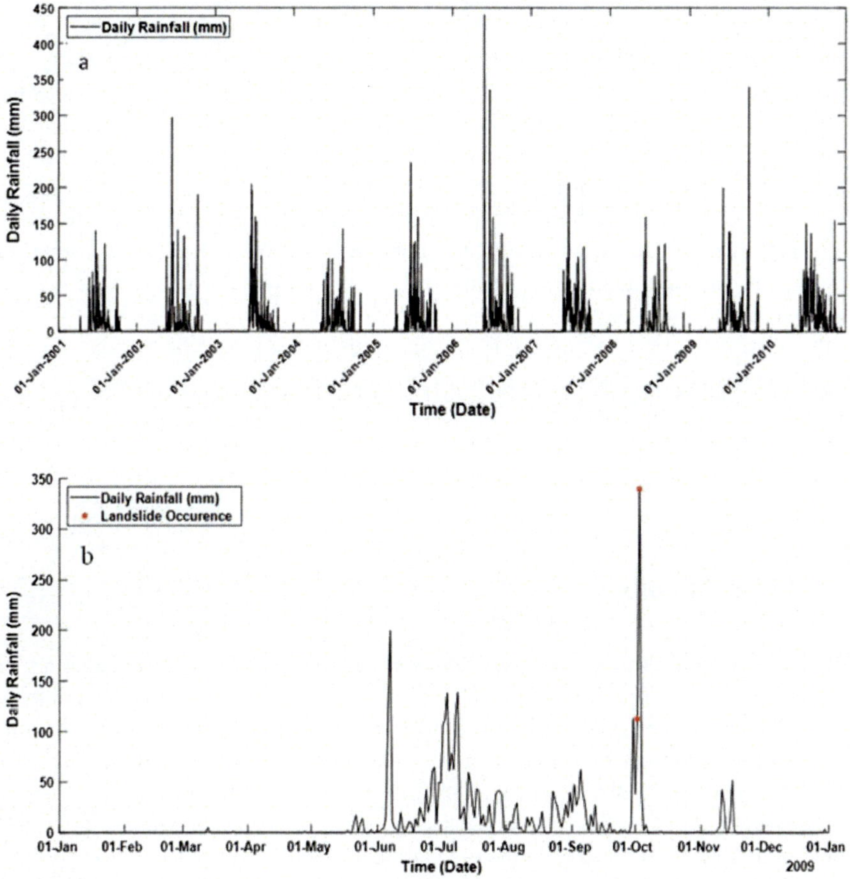

Fig. 6.4 **a** Daily rainfall in Karwar District shown from January 2001 to December, 2010. **b** Daily rainfall in Karwar District shown from 01st January 2009 to 31st December, 2009 with landslide events marked with red asterisks

observed, the rainfall for 2nd October, 2009 was not nearly high enough to account for the landslides in the area. This indicates a significant relation between antecedent rainfall and landslide initiation.

To investigate the relationship between antecedent rainfall and landslide initiation, the total rainfall for all events (E) and the corresponding duration (D) was derived from daily rainfall records and is presented in Table 6.1. All the extreme rainfall events and durations from 2001 to 2010 along with the corresponding E_{MAP} are listed in Table 6.1. Cumulative rainfall (E) is defined as the total rainfall for an uninterrupted series of rainy days. This measure i.e. E combined with the duration of event provides useful insights into the mechanism of Landslides at Karwar and rainfall intensity/duration. When the relationship between E and D is plotted as in Fig. 6.5, it is clear that the rainfall events resulting in landslide occurrence involved a minimum of 11 days of continuous rainfall. Also, the minimum threshold for cumulative event rainfall (E) for landslide initiation is around 260 mm. On normalizing E with the Mean Annual Precipitation (MAP) of Karwar district i.e. 3277 mm, a minimum E_{MAP} of 0.08 is obtained. Therefore, for events having E_{MAP} values greater than or equal to 0.08 and having duration of more than 11 days, a landslide incidence is very likely. The events on 10-07-2001 and 29-07-2001 are the only anomalies to the above derived threshold. However, these can be explained due to the fact the effect of human induced disturbances such as quarrying, construction etc. during the early years of 2000s was not as severe as during the latter part of the decade.

Thus, this study indicates that for the coastal region of Western Ghats of Karnataka, both peak daily rainfall as well as antecedent rainfall is important factors in landslide initiation. A minimum of 11 days of continuous rainfall with cumulative rainfall above 260 mm is necessary to trigger landslides and can be attributed to the combination of saturation of soil caused by antecedent precipitation and short-term high intensity rainfall events. This method provides an early warning signal to local administrative authorities and can be used for advanced preparedness and relief actions.

6.5 Conclusions

The study shows a strong interdependence between landslide triggering and rainfall pattern. The landslide incidence is a direct function of the magnitude of the event rainfall depth. Events having E_{MAP} value above the threshold of 0.08 (above 260 mm) and continuous rainfall duration of 11 days or more caused triggering of landslide in Karwar that particular year. Thus, the rainfall events with intensity less than the minimum threshold value did not cause triggering of landslide whereas landslides occurred when the twin thresholds of rainfall and duration are crossed. There were some rainfall events for which the landslide occurrence criteria was true but landslide did not occur. This shows that the slope was stable enough preventing slope failure even though there was required amount of rainfall for landslide initiation. The possible cause for the triggering of Kadwad landslide is a combination of saturation

Table 6.1 Cumulative rainfall and their Durations along with corresponding E_{MAP}

Date	Duration of event, *D*		Cumulative rainfall event, E (in mm)	E_{MAP}
	(in hours)	(in days)		
Events with no landslide				
27-05-2001	192.0	8	249.6	0.08
11-06-2001	120.0	5	157.3	0.05
24-06-2001	168.0	7	213.9	0.07
02-09-2001	384.0	16	177.2	0.05
15-10-2001	216.0	9	77.4	0.02
22-05-2002	72.0	3	127.2	0.04
07-07-2002	240.0	10	75.4	0.02
20-07-2002	288.0	12	223.2	0.07
30-07-2002	216.0	9	130.8	0.04
07-08-2002	144.0	6	104.9	0.03
11-08-2002	72.0	3	264.5	0.08
27-08-2002	288.0	12	135.8	0.04
15-10-2002	168.0	7	293.9	0.09
17-06-2003	192.0	8	358.7	0.11
27-08-2003	384.0	16	238.7	0.07
24-10-2003	96.0	4	47.1	0.01
18-05-2004	336.0	14	195.7	0.06
19-06-2004	168.0	7	75.7	0.02
17-07-2004	216.0	9	241.2	0.07
13-08-2004	48.0	2	171.9	0.05
05-10-2004	336.0	14	244.1	0.07
19-06-2005	312.0	13	220.1	0.07
05-07-2005	216.0	9	152	0.05
27-08-2005	312.0	13	145.7	0.04
27-09-2005	192.0	8	195.8	0.06
29-05-2006	168.0	7	36.5	0.01
28-07-2006	192.0	8	152.6	0.05
06-10-2006	120.0	5	182.5	0.06
04-06-2007	168.0	7	178.3	0.05
23-07-2007	240.0	10	145.4	0.04
16-08-2007	192.0	8	96.3	0.03
25-09-2007	264.0	11	171.7	0.05
05-10-2007	168.0	7	112.2	0.03
05-06-2008	456.0	19	141.8	0.04

(continued)

Table 6.1 (continued)

Date	Duration of event, *D*		Cumulative rainfall event, E (in mm)	E_{MAP}
	(in hours)	(in days)		
22-09-2008	264.0	11	152.6	0.05
27-05-2009	216.0	9	52.8	0.02
27-06-2009	192.0	8	165.6	0.05
26-07-2009	144.0	6	81.2	0.02
31-08-2009	216.0	9	174	0.05
03-06-2010	72.0	3	53.4	0.02
17-06-2010	144.0	6	204.2	0.06
09-10-2010	144.0	6	33.1	0.01
21-10-2010	168.0	7	99	0.03
Events with landslide				
02-10-2009	264.0	11	280.15	0.09
03-10-2009	288.0	12	620.1	0.19
Anomalous events				
10-07-2001	360.0	15	521.2	0.16
29-07-2001	312.0	13	306.8	0.09

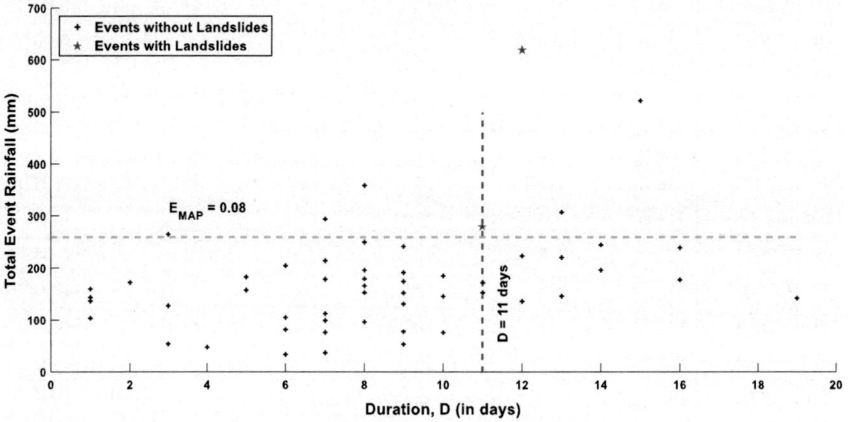

Fig. 6.5 Cumulative event rainfall (*E*) versus event duration (*D*)

of soil caused by antecedent precipitation and short term high intensity rainfall along with human interventions.

Acknowledgements The authors greatly acknowledge National Institute of Technology, Surathkal, Karnataka for providing financial assistance and facilities for carrying out the research work.

References

Aleotti P (2004) A warning system for rainfall-induced shallow failures. Eng GEol 73(3):247–265
Bai S, Wang J, Thiebes B, Cheng C, Yang Y (2014) Analysis of the relationship of landslide occurrence with rainfall: a case study of Wudu County, China. Arab J Geosci 7(4):1277–1285
Brand EW, Premchitt J, Phillipson HB (1984) Relationship between rainfall and landslides in Hong Kong. In: Proceedings of the 4th international symposium on landslides. BiTech Publishers, Vancouver, vol 1, pp 377–384
Caine N (1980) The rainfall intensity-duration control of shallow landslides and debris flows. Geografiskaannaler: Ser A, Phys Geogra 62(1–2):23–27
Cannon SH, Ellen SD (1985) Rainfall conditions for abundant debris avalanches in the San Francisco Bay region, California. Calif Geol 38(12):267–272
Cardinali M, Galli M, Guzzetti F, Ardizzone F, Reichenbach P, Bartoccini P (2006) Rainfall induced landslides in December 2004 in south-western Umbria, central Italy: types, extent, damage and risk assessment. Nat Hazards Earth Syst Sci 6(2):237–260
Claessens L, Schoorl JM, Veldkamp A (2007) Modelling the location of shallow landslides and their effects on landscape dynamics in large watersheds: an application for Northern New Zealand. Geomorphology 87(1–2):16–27
Crosta GB, Frattini P (2001) Rainfall thresholds for triggering soil slips and debris flow. In: Proceedings of the 2nd EGS Plinius conference on mediterranean storms: Publication CNR GNDCI, vol 2547, pp 463–487
De Vita P (2000) Phenomena of instability of pyroclastic coverings of the Monti Lattari, Sarno and Salerno (Campania) and analysis of the pluviometric determinants. Quaderni Appl Geol 7(2):213–235
Gabet EJ, Burbank DW, Putkonen JK, Pratt-Sitaula BA, Ojha T (2004) Rainfall thresholds for landsliding in the Himalayas of Nepal. Geomorphology 63(3–4):131–143
Glade T, Crozier M, Smith P (2000) Applying probability determination to refine landslide-triggering rainfall thresholds using an empirical Antecedent Daily Rainfall Model. Pure Appl Geophys 157(6–8):1059–1079
Govi M, Mortara G, Sorzana PF (1985) Hydrological events and landslides. GeologiaApplicata e Idrogeologia 20:359–375
Guzzetti F, Peruccacci S, Rossi M, Stark CP (2007) Rainfall thresholds for the initiation of landslides in central and southern Europe. Meteorol Atmos Phys 98(3–4):239–267
Hegde VS, Krishnaprasad PA, Shalini R, Nilamwar D, Tejaswini B, Girish KH, Malewadi CS (2014) Land-slide hazards of October 2009 at Karwar, Karnataka: A lesson for planning developmental activities in the tropical Ghat regions. Int J Earth Sci Eng 7:260–268
Heyerdahl H, Harbitz CB, Domaas U, Sandersen F, Tronstad K, Nowacki F, Engen A, Kjekstad O, Devoli G, Buezo SG, Diaz MR (2003) Rainfall induced lahars in volcanic debris in Nicaragua and El Salvador: practical mitigation. In: Proceedings of international conference on fast slope movements—prediction and prevention for risk mitigation, IC-FSM2003. Patron Pub, Naples, pp 275–282
Jakob M, Weatherly H (2003) A hydroclimatic threshold for landslide initiation on the North Shore Mountains of Vancouver, British Columbia. Geomorphology 54(3–4):137–156
Kim SK, Hong WP, Kim YM (1991) Prediction of rainfall-triggered landslides in Korea. In: Bell DH (ed) Landslides, 2nd edn. A.A. Balkema, Rotterdam, pp 989–994
Lan H, Zhou CH, Lee CF, Wang S, Wu FQ (2003) "Rainfall-induced landslide stability analysis in response to transient pore pressure-A case study of natural terrain landslide in Hong Kong" . Science in China Ser E Technol Sci 46:52–68
Peruccacci S, Brunetti MT, Gariano SL, Melillo M, Rossi M, Guzzetti F (2017) Rainfall thresholds for possible landslide occurrence in Italy. Geomorphology 290:39–57
Rosi A, Segoni S, Catani F, Casagli N (2012) Statistical and environmental analyses for the definition of a regional rainfall threshold system for landslide triggering in Tuscany (Italy). J Geog Sci 22(4):617–629

Sengupta A, Gupta S, Anbarasu K (2010) Rainfall thresholds for the initiation of landslide at LantaKhola in north Sikkim, India. Nat Hazards 52(1):31–42

Terlien MT (1998) The determination of statistical and deterministic hydrological landslide-triggering thresholds. Environ Geol 35(2–3):124–130

Tsai TL (2008) The influence of rainstorm pattern on shallow landslide. Environ Geol 53(7):1563–1569

Chapter 7
Optimal Cropping Pattern of Kulsi River Basin, Assam, India Using Simulation and Linear Programming Model

Jyotismita Taye, Bibhash Sarma, and Abhijit D. Lade

Abstract Irrigation potential of an area can be benefitted if we thrust on maximizing crop production by suitable allocation of limited resources to get optimal productivity. The Kulsi River Basin, located in the Southern bank of the Brahmaputra River, is the area of study. The gross command area and net irrigated area in the basin are 37,908 Ha and 23,882 Ha, respectively. The river basin is deficit of regular irrigation facilities, and only traditional cropping pattern (rain-fed conditions only) has been followed by the cultivators of the area. Hence, to promote the areas potential, the Kulsi Multipurpose Project Authority of Brahmaputra Board has proposed a multipurpose reservoir for the generation of hydropower, flood control, and irrigation. The capacity of the reservoir is 525.64 MCM, plant capacity is 55 MW, tail water level is 63 m, and dead storage of the reservoir is 85.86 MCM. Keeping in mind the benefits to be received by the proposed reservoir, the objective of the paper is mainly emphasized to compute the crop water requirement of the Kulsi Basin and to formulate a simulation and linear programming (LP) model to find the best possible cropping arrangement in order to maximize the net benefit of the crops.

Keywords Irrigation potential · Optimal productivity · Crop water requirement · Cropping pattern · Simulation and linear programming model

7.1 Introduction

With the increasing demand due to residents' growth and rapid industrialization, water is becoming valued and scarce. The rising demand and scarcity of water makes it important to use the available water in the most economical way. Crops grow with an adequate amount of water. Proper planning of crops to be grown in an area can be

J. Taye (✉) · A. D. Lade
Indian Institute of Technology, Guwahati, India
e-mail: taye176104022@iitg.ac.in

B. Sarma
Assam Engineering College, Guwahati, India

R. Jha et al. (eds.), *Water Resources Management and Reservoir Operation*, Water Science and Technology Library 107,
https://doi.org/10.1007/978-3-030-79400-2_7

assured with sufficient amount of water present there. Higher crop productivity can only be attained by proper soil, water, and crop management in the areas with assured irrigation. The key to effective irrigation water management lays in proper estimation of crop water requirement, which are primarily based on cropping pattern, rainfall in the area, and other climatology factors. The objective of the paper is emphasized to study the water requirements of different crops and to obtain the most favorable cropping pattern in order to maximize the net profit of crops using simulation and LP model.

Planning of optimal cropping pattern needs evaluation of crop water requirements (CWR) of individual crops. Each crop has different water requirements. Food and Agriculture Organization of the United Nations, FAO (Doorenbos 1977) has given procedures to calculate crop water requirements of crops by a software called CROPWAT. It is a tool to carry out standard calculations for reference evapotranspiration, crop water requirements and crop irrigation requirements, and more specifically the design and management of irrigation schemes (Karamouz et al. 2010). CROPWAT has been used by various researchers to evaluate the water requirements of crops. Kuo and Liu (2002) presented a CROPWAT model where they evaluated the water requirements of crops in Taiwan. Similar work has been carried out by Kia (2013) in Iraq, Saravanan and Saravanan (2014) in Tamil Nadu, India, and Surendran et al. (2015) in Palakkad district, Kerala, India. The Kulsi River having the potential of supplying irrigation water needs proper planning for the distribution of water. Sarma et al. (2013) have shown how optimization and simulation can be combined for sizing Kulsi reservoir system. Optimal cropping pattern and production of food crops with maximum profit are important information for irrigation planning which is achieved by optimization methods. In this study, linear programming optimization model is used. Many researchers have carried out various works with the use of linear programming. Singh et al. (2001), Vedula (1992), Raman et al. (1992), Mainuddin et al. (1997), Sarker et al. (1997), Kangrang et al. (2010) made use of linear programming technique for optimum utilization of resources (land and water) considering various resource and requirement constraint. The objective of the paper includes:

- To compute crop water requirements of Kulsi Basin
- To formulate a simulation and linear programming model to find
  - (a) The power potential of the proposed Kulsi Multipurpose Project
  - b) The irrigation potential of the proposed Kulsi Multipurpose Project
  - (c) The optimal cropping pattern for maximization of net benefit
  - (d) The area under different crops for maximization of net benefit.

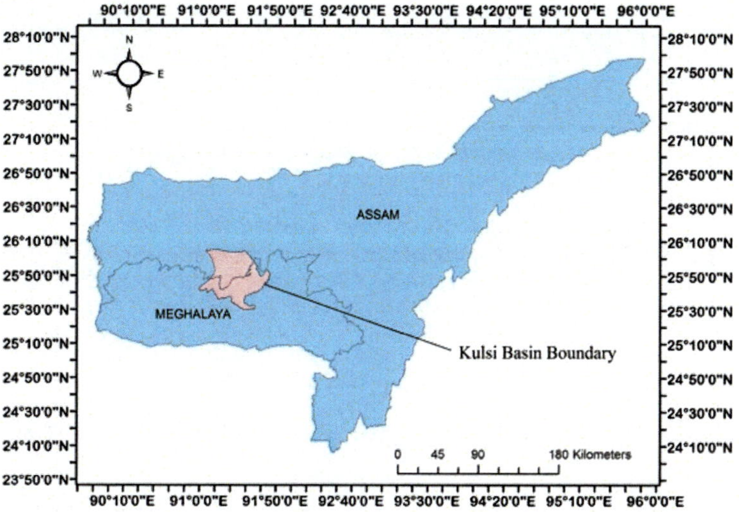

Fig. 7.1 Study map of the Kulsi River Basin

7.2 The Study Area

The Kulsi River Basin is a tributary (south bank) of the river Brahmaputra situated in Kulsi-Deosila sub-basin. The river is formed mainly by three tributaries, Khri, Krishniya, and Umsiri. They all emerge from west Khasi Hills ranges (Fig. 7.1).

It covers an overall area of 3770 km^2 within Assam and Meghalaya. Out of the total catchment areas, 685 km^2 is plain catchment in Assam and 3085 km^2 is hill catchment in Meghalaya and Assam. The total length of Kulsi is about 196 km from its source to its outfall.

7.3 Methods and Materials

7.3.1 Crop Water Requirement

The depth of water consumptively used by a crop (plus unavoidable irrigation application losses) may be defined as crop water requirement. The Food and Agriculture Organization of the United Nations (FAO) has given procedures to calculate crop water requirements of crops by a software called CROPWAT. The nature and stages of growth of the crops including the environmental conditions influence the water requirement of crops. Figure 7.2 shows the change in crop coefficient during different growth stages of crop. Higher temperatures will increase transpiration from vegetation, boosting water requirements for both rain-fed and irrigated crops.

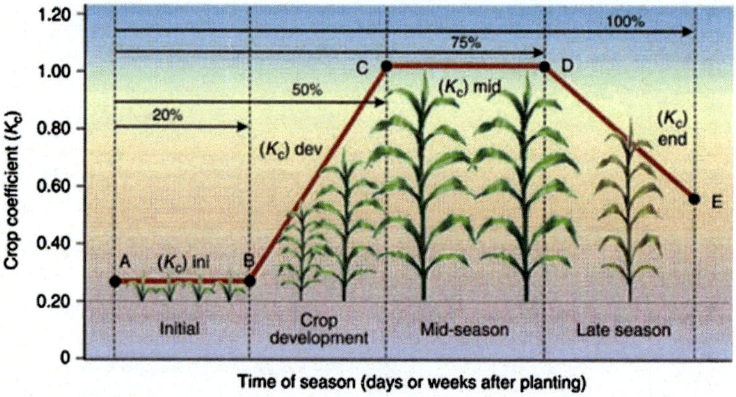

Fig. 7.2 Schematic representation of crop coefficient during different growth stages of crop (Courtesy: Pokorny 2018)

In this study, computation of crop water requirement has been carried out in CROPWAT 8.0 by calling up successively the suitable climatic and rainfall data, together with the crop data and corresponding planting dates of crops. For the calculation of crop water requirement, the model requires the following:

- Reference crop evapo-transpiration (ET_o): The $ET_o$of Guwahati has been calculated by the FAO Penman–Monteith method in CROPWAT 8.0. For the calculation of ET_o, the FAO Penman–Monteith method needs the records of radiation, air temperature, air humidity, wind speed, and sunshine hours of the area selected.
- Rainfall data: The rainfall data have been collected from the Meteorological center at Gopinath Bordoloi International Airport site maintained by IMD for the year 1951 to 2000 to find out the effective rainfall. Effective rainfall is the total amount of rainwater useful for meeting the water need of the crops.
- Crop data: This part of calculation in CROPWAT requires the value of crop coefficient (K_c) Table (7.1), growth period of crops, planting date, rooting depth, yield response factor, critical depletion of crop, and crop height.

In the absence of regular irrigation facilities in the area, only traditional cropping pattern has been followed till now by the cultivators of the area and the crops are grown in rain-fed condition only. The existing cropping intensity in the area is found to be 146.56%.

7.3.2 The Simulation Model

Simulation models are considered as a valuable tool for understanding the process of a system including the operators' knowledge and decision (Khadr and Elshemy 2017). In this paper, in order to observe the behavior of the proposed reservoir by

Table 7.1 Crop coefficient (K_c) of some crops during different stages of crop growth

Crops	Initial stage	Crop development stage	Mid-season stage	Late season stage
Rice	1.05	1.2	0.8	1
Jute	0.4	1.15	0.5	2
Onion	0.5	0.75	1.05	0.85
Potato	0.45	0.75	1.15	0.85
Pea	0.45	0.8	1.15	1.05
Radish	0.45	0.6	0.9	0.9
Maize	0.4	0.8	1.15	0.7
Sugarcane	0.3	1.2	0.5	2
Lentil/Pulses	0.45	0.75	1.1	0.65

Kulsi Multipurpose Project, a simulation model is prepared. With 40 years of input data, the simulation model was run. The simulation model is assumed to have an initial storage of the reservoir as equal to dead storage. May is considered as the start of each year. The best suited equations for elevation versus capacity (Fig. 7.3) and area versus capacity (Fig. 7.4) are obtained.

Collection of data and processing are essential steps to formulate the model. The following are the hydrological data of the River Kulsi collected from Brahmaputra Board to formulate the simulation model:

- Inflow data at monthly time step (Table 7.2).
- Mean monthly evaporation data.
- Area–capacity–elevation characteristics of the reservoir.
- Irrigation demand.

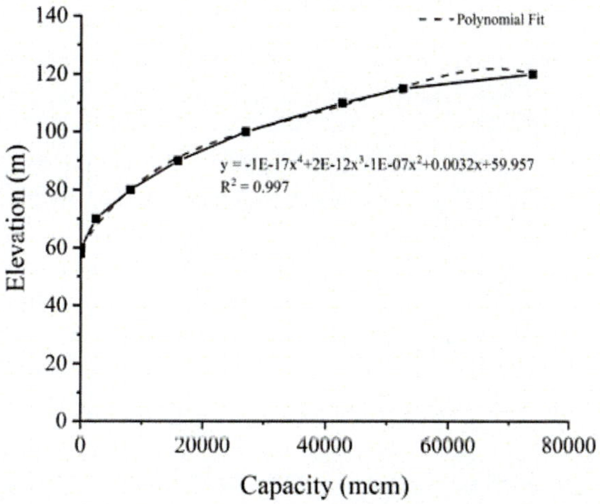

Fig. 7.3 The elevation versus capacity curve with best suited equation

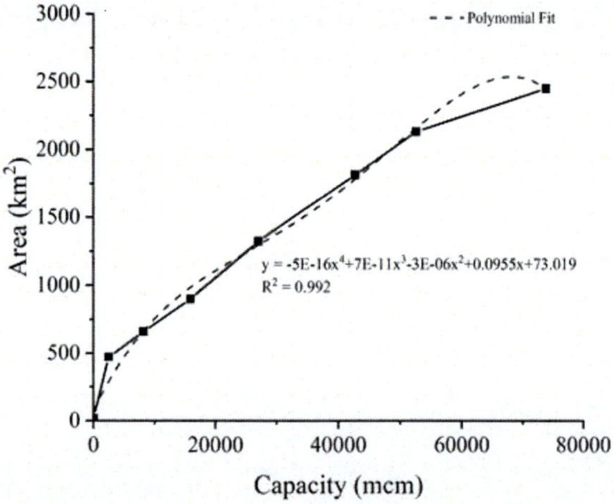

Fig. 7.4 The area versus capacity curve with best suited equation

Table 7.2 Average inflow (MCM) at each month of the year

Months	Average inflow (MCM)	Months	Average inflow (MCM)
May	47.46	November	52.3
June	94.87	December	36.53
July	150.48	January	25.57
August	132.85	February	21.56
September	117.43	March	18.81
October	93.97	April	29.77

The elevation–area–capacity values have been worked out from reservoir area maps developed by the Kulsi Multipurpose Project Authority. Original elevation–area–capacity values have been considered for the purpose of the study.

The full reservoir level (FRL) for the project has been fixed at El 115.00 m. The elevation–capacity and area-capacity curves are generated using the data given in Table 7.3.

The irrigation demand considered for the purpose of the study is taken from the computation of crop water requirement. The irrigation demand of all the months is shown in Fig. 7.5.

As proposed by the Brahmaputra Board, the reservoir capacity is taken as 525.64 MCM, plant capacity as 55 MW, and tail water level as 63 m. Initially, firm power is assumed as 0.5 MW. From the data available and using it with the help of the simulation model, the power and irrigation potential are obtained by satisfying the minimum target reliability.

Table 7.3 The elevation–area–capacity values of the reservoir of the Kulsi Multipurpose Project

Elevation (m)	Area (ha)	Capacity (ha m)
58	0	0
60	18	18
70	471	2483
80	660	8118
90	899	15,913
100	1323	27,023
110	1813	42,703
115	2131.5	52,564.25
120	2450	73,879.25

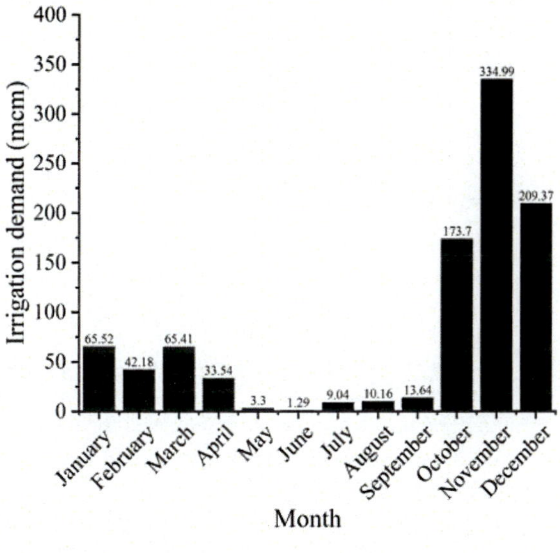

Fig. 7.5 Monthly distribution of irrigation demand (MCM)

7.3.3 *Linear Programming Model*

An important factor of agricultural benefit is its planning of cropping pattern. Land, water, labor, and capital are some of the resources on which cropping pattern depends (Sarker and Quaddus 2002). Optimal cropping pattern and production of food crops with maximum profit is important information for irrigation planning using optimization methods. In this study, the LP model prepared using Lingo is used to formulate the optimal cropping pattern in the basin. The model is prepared with the aim of maximization of net benefit of the crops in a given area. The objective function in the model is formulated to determine the cropping pattern (optimal) in the area of study that will maximize the net benefit of the crops. The decision variable in the formulated LP model is the area under the crops (X_k). The constraints that are considered are area constraint, water constraint, and affinity constraint. The prepared LP model is run once by considering the affinity constraint and once without considering affinity constraint.

Objective Function:

- Maximization of net benefit
  Maximize Z
  Subject to

$$Z = \sum_{k-1}^{\mathrm{NC}} b_k C_k X_k$$

where k denotes particular crop from number 1 to NC.

NC	Number of crops
b_k	Net benefit of crop k per ton
C_k	Crop yield in ton per hectare of crop k
X_k	Area under crop k in hectares

Constraints:

The resource availability constraints which the model should satisfy while achieving the desired objectives are:

- Land availability constraints
  The total area assigned to different crops in a particular season cannot surpass the culturable command area (CCA) of the area. Thus, the area assigned to Kharif and perennial crops should not exceed the CCA in summer and the area assigned to Rabi and perennial crops also should not exceed CCA in winter. This can be expressed as:

$$\sum_{k} X_k \leq \mathrm{CCA}$$

where X_k = Area under crop k in hectares

CCA = Culturable command area of the region

- Water availability constraints
  The water requirement of the crops should be fully met during the entire crop period. It should not exceed the available water. The total irrigation demand of a crop k for a given growth period of the crop is taken as the water available for crop. It is expressed as:

$$\sum_{k=1}^{NC} W_{k,t} X_k \leq O_t$$

where $W_{k,t}$ = Water depth of crop (crop water requirement) k at time t

O_t = Total irrigation demand of crop k at time t

- Affinity constraint
  Affinity constraint is considered to bind the area under the crops within their upper limit or lower limit. It is expressed as:

$$\mathrm{LL}_k \leq X_k \leq \mathrm{UL}_k$$

where LL_k = Lower limit of area for crop k in hectares

UL_k = Upper limit of area for crop k in hectares

7.4 Results and Discussions

(1) Crop water requirement (CWR)

The study of crop water requirement is carried out by CROPWAT 8.0. Before the calculation of crop water requirement of each crop, it is necessary to calculate the reference evapo-transpiration (ET_o) and the effective rainfall of the area. The evapo-transpiration rate from a reference surface (a hypothetical grass reference crop with an assumed crop height), not short of water, is called the reference crop evapo-transpiration or reference evapo-transpiration and is denoted as ET_o (Allen et al. 1998). The FAO Penman–Monteith method is suggested as the only method for determining ET_o. Table 7.4 shows the ET_o measured using CROPWAT.

Effective rainfall is the total amount of rainwater useful for meeting the water need of the crops. The rainfall data have been collected from the Meteorological center at Gopinath Bordoloi International Airport site maintained by IMD for the year 1951 to 2000, which has been considered for the study of the basin. For the study of crop water requirement, the effective rainfall is calculated by CROPWAT. The FAO CROPWAT 8.0 uses the USDA Soil Conservation Service method by default to calculate the effective rainfall. The only input here is to provide the average rainfall data of every month in millimeters. Table 7.5 shows the effective rainfall measured.

The crop data and soil data are entered into the CROPWAT to find the crop water requirements of each crop. The soil data and crop data are different for different crops, and it is to be found locally. The result shows that the CWR for onion (306.9 mm/year) and garlic (306.9 mm/year) is the maximum in the Rabi season. Cauliflower and radish require less water than the other crops. The CWR for sugarcane, which is an annual crop, is 352.8 mm/year. The maximum CWR is for summer paddy, i.e., local boro (432.4 mm/year) and minimum for radish (5.9 mm/year). The CWR is

Table 7.4 Reference crop evapo-transpiration (ET_o) per month

Month	Min temp	Max temp	Humidity	Wind speed	Sunshine	Radiation	ET_o
	°C	°C	%	km/day	Hours	$MJ/m^2/day$	mm/day
January	10.2	23.6	89	1	10.4	17.5	2.17
February	11.9	26.2	83	1	11.3	21.1	2.89
March	15.8	30	64	3	11.6	24.4	3.78
April	20	31.2	72	3	12.4	27.7	4.99
May	22.6	31.1	77	2	13.2	29.8	5.71
June	25	31.7	82	2	13.4	30.1	6.13
July	25.5	31.8	86	2	13.4	30	6.22
August	25.5	32.1	83	1	13	28.7	5.9
September	24.6	31.6	86	1	12.2	25.8	5.2
October	21.9	30.2	86	1	11.3	21.8	4.06
November	16.6	27.6	85	1	10.5	18.1	2.85
December	11.7	24.6	90	1	10.3	16.6	2.19
Average	19.3	29.3	82	2	11.9	24.3	4.34

Table 7.5 Effective rainfall (mm) at each month of the year

Months	Rain (mm)	Effective rainfall (mm)
January	11.8	11.6
February	17.2	16.7
March	55.1	50.2
April	147.1	112.5
May	248.9	149.8
June	316.8	156.7
July	351.2	160.1
August	269.4	151.9
September	186.9	131
October	90.9	77.7
November	17.8	17.3
December	6.8	6.7
Total	1719.9	1042.3

maximum in the November month (1089.2 mm) and minimum in the month of June (36.8 mm).

(2) **Simulation model**

The application of simulation model is started with the assumption that the initial storage in the reservoir is taken equal to the dead storage 85.86 MCM. After developing and solving the model, the power potential and irrigation potential of the reservoir are obtained. The result shows that the project has a potential of supplying 934 MCM of water annually for irrigation needs. Considering the annual irrigation demand as 934 MCM, it was found that a firm power of 0.7 MW can be produced with the target reliability, i.e., irrigation reliability of 75.05% and power reliability of 98.93%.

(3) **Linear programming model**

The LP model is prepared using Lingo. The model is prepared with the objective of maximization of net benefit. The model is run once considering affinity constraint and once without affinity constraint and the results are observed for each run. When considering affinity constraints, the model takes 10% of the area of proposed cropping pattern as minimum area for maximization of net benefit. The result shows that for maximization of net benefit without considering affinity constraint, the linear programming model provides a net benefit of ₹19,105.6 lakhs considering only one crop (sugarcane), while considering affinity constraint the model provides a net benefit of ₹ 19,624.7465 lakhs (Table 7.6), where sugarcane takes the maximum area among the other crops. Moreover, an optimal cropping pattern with variety of crop distribution within the area is proposed with the presence of affinity constraint.

7.5 Conclusion

In this paper, an attempt has been made to plan an optimal cropping pattern for Kulsi River Basin in Assam, India. The first study carried out was to find out the crop water requirements of the existing crops in the basin. The second study was to formulate a simulation model to study the behavior of the reservoir as proposed by the Kulsi Multipurpose Project Authority. The third study was to find out the optimal cropping pattern for maximization of net benefit using linear programming model. Crop water requirement is an important parameter to study as it allows us to know how effective the crop root zone is in storing water. Maximizing the net benefit of the crops will benefit the cultivators in choosing the appropriate crops. This optimal cropping pattern may serve as a guideline to adopt a new cropping pattern in this area or any other similar area nearby.

Table 7.6 Crop area distribution to maximize net benefit in presence of affinity constraint

Crop area distribution to maximize net benefit (Presence of affinity constraint)					
Crops	Area under crops for maximizing net benefit (ha)	Yield (MT/ha)	Total yield in MT	Rate per MT	Benefit (₹ in lakh)
Summer paddy local paddy	70	4	280	8000	22.4
Summer paddy local early ahu	24	2.5	60	8000	4.8
Autumn paddy local regular ahu	35	2.4	84	7000	5.88
Bottle gourd	10	15	150	8000	12
Jute	229.3	1.5	343.95	15,000	51.5925
Winter paddy local	380	3	1140	7500	85.5
Winter paddy HYV Ranjit Bahadur	1200	4	4800	7000	336
Black and green gram	157.5	0.6	94.5	30,000	28.35
Sesame	109.2	0.8	87.36	40,000	34.944
Potato	1964.5	8	15,716	10,000	1571.6
Maize	85.7	3	257.1	10,000	25.71
Onion	71.9	8	575.2	5000	28.76
Garlic	92.9	3	278.7	30,000	83.61
Sugarcane	21,667	40	866,680	2000	17,333.6
Total benefit					19,624.7465

References

Allen RG, Pereira LS, Raes D, Smith M (1998) Crop evapotranspiration-Guidelines for computing crop water requirements—FAO Irrigation and drainage paper 56. FAO, Rome 300(9):D05109

Doorenbos J (1977) Guidelines for predicting crop water requirements. FAO Irrigation and Drainage Paper, vol 24, pp 1–179

Kangrang A, Compliew S (2010) An application of linear programming model for planning dry-seasonal Irrigation system. Trends Appl Sci Res 5(1):64–70

Karamouz M, Semsaryazdi M, Ahmadi B, Ahmadi A (2010) Climate change impacts on crop water requirements: a case study. In: 1st IWA Malaysia young water professionals conference, international water association, pp 1–11

Khadr M, Elshemy M (2017) Data-driven modeling for water quality prediction case study: the drains system associated with Manzala Lake, Egypt Ain Shams Eng J 8(4):549–557

Kia DR (2013) Water requirements for major crops in different agro-climatic zones of Iraqi Kurdistan using by CROPWAT 8.0. IOSR J Agric Veter Sci 6(3):30–36
Kuo SF, Liu CW (2002) Simulation and optimization model for irrigation planning and management. Hydrolog Proc 17(15):3141–3159
Mainuddin M, Gupta AD, Onta PR (1997) Optimal crop planning model for an existing groundwater irrigation project in Thailand. Agric Water Manage 33(1):43–62
Pokorny J (2018) Evapotranspiration. Encyclopedia of Ecology. Ref Module Earth Syst Environ Sci 2:292–303
Raman H, Mohan S, Rangacharya NCV (1992) Decision support system for crop planning during droughts. J Irrigat Drainage Eng 118(2):229–241
Saravanan K, Saravanan R (2014) Determination of water requirements of main crops in the tank irrigation command area using CROPWAT 8.0. Int J Interdisc Multidiscipl Stud 1(5):266–272
Sarker RA, Quaddus MA (2002) Modeling a nationwide crop planning problem using a multiple criteria decision making tool. Comput Ind Eng 42(2–4):541–553
Sarker RA, Talukdar S, Haque AFMA (1997) Determination of optimum crop mix for crop cultivation in Bangladesh. Appl Math Model 21(10):621–632
Sarma B, Das KK, Patwari NN (2013) Sizing of Kulsi reservoir system using non-linear optimization and simulation. In: Proceedings of Assam international water conference - 2013, 21–22 February, 2013, Guwahati
Singh DK, Jaiswal CS, Reddy KS, Singh RM, Bhandarkar DM (2001) Optimal cropping pattern in a canal command area. Agric Water Manage 50(1):1–8
Surendran U, Sushanth CM, Mammen G, Joseph EJ (2015) Modelling the crop water requirement using FAO-CROPWAT and assessment of water resources for sustainable water resource management: a case study in Palakkad district of humid tropical Kerala, India. Aquatic Procedia, 4, 1211–1219
Vedula S, Mujumdar PP (1992) Optimal reservoir operation for irrigation of multiple crops. Water Res Res 28(1):1–9

Chapter 8
Comparison of Flux Footprint Models to a Mixed Fetch Heterogeneous Cropland System

Shweta Kumari and K. B. V. N. Phanindra

Abstract Flux footprint models delineate the source area of micrometeorological fluxes as measured by the eddy covariance (EC) system. An accurate knowledge of flux footprint is crucial for effective interpretation of measured fluxes and upscaling to a regional scale with the help of satellite data. Flux footprint models are derived based on the assumption that "measured fluxes are originated from a homogeneous cropland system." For small and fragmented land-holdings of the Indian agricultural system, this assumption is often violated, hence questioning the applicability of existing models. The objective of this paper is to evaluate the performance of various footprint models to a mixed fetch (heterogeneous) cropland system. Field-scale experiments were conducted in continuous cotton (C3) sugarcane (C4) plots located in Nandikandi village, Telangana, India. Two low height EC towers capturing homogeneous fluxes from individual fields and one tall EC tower capturing heterogeneous fluxes from the combinations were used in the study. Half hourly fluxes are calculated as the covariance between vertical wind velocity and scalar concentration of interest (water vapor, air temperature, or carbon dioxide). Three analytical models (viz. Hsieh, Schuepp, and Kormann and Meixner) were applied along with the Lagrangian particle dispersion model, considering unstable (daytime) and stable (night time) flux measurements. Our results conclude that, for homogeneous croplands, the Schuepp model performs best during unstable atmospheric condition and Hsieh model performs best in stable atmospheric conditions. Hseih model is suitable in representing the source area fluxes in a mixed fetch condition for all atmospheric stability conditions.

Keywords Flux footprint · Eddy covariance · Analytical model · Lagrangian stochastic model · Cropland

S. Kumari (✉) · K. B. V. N. Phanindra
Department of Civil Engineering, Indian Institute of Technology Hyderabad, Hyderabad 502284, India

R. Jha et al. (eds.), *Water Resources Management and Reservoir Operation*, Water Science and Technology Library 107,
https://doi.org/10.1007/978-3-030-79400-2_8

8.1 Introduction

Flux footprint represents the upwind region from where the atmospheric fluxes measured by an instrument are generated. Several atmospheric (atmospheric stability, surface roughness, and measurement height), surface (homogenous, heterogeneous), and phenological properties can influence the shape and extent of footprint (Kolari et al. 2004). Eddy covariance (EC) methodology allows for the quantification of the fluxes of interest (water vapor, air temperature, carbon dioxide, and methane) by co-variating with vertical wind speed, at high frequency. A footprint model is a mathematical representation of the spatial distribution of flux sources and the corresponding magnitude with respect to the measurement location (Arriga et al. 2017). Footprint models help in visualizing the spatial distribution of source area fluxes as measured by the EC flux towers. Methods for modeling flux footprint include: (a) analytical models (computationally least expensive, solves advection–diffusion equation, and simple to use), (b) Lagrangian particle models (accounts for three-dimensional turbulent diffusion and non-Gaussian inhomogeneous turbulence), and (c) large-eddy simulations (address spatial heterogeneity and non-topography explicitly).

Analytical models consider atmospheric and crop phenological characteristics to develop flux footprint and are largely constrained by a number of assumptions. These include (a) horizontal homogeneity, (b) stationarity, and (c) zero mean vertical wind speed during the averaging period. These assumptions are a major limitations of the Eddy covariance system application to real world fields and question the accuracy of the predicted flux footprint.

One of the primary assumptions, on which the analytical models were built is that the contributing source area to the flux measurements has homogeneous cropland and surface roughness characteristics. This assumption can be violated under two conditions: (i) for small and fragmented croplands, wherein multiple crop fields exist within the fetch of a tower, and (ii) stable atmospheric conditions with calm wind, where fetch is considerably large. For heterogeneous surfaces (that are commonly seen in the Indian agricultural setting), there is a dependency of measured signal on how strongly a certain part of the surface influences the sensor, and thus the location and size of its footprint do affect the measured signal.

Applicability of flux footprint models to either homogeneous or heterogeneous croplands of the Indian sub-continent was never explored in literature. This could be due to the non-availability of flux datasets for model validation. The objective of this work is to evaluate the performance of analytical models applied to homogeneous and heterogeneous cropland systems considering daytime unstable and nighttime stable atmospheric conditions. Data from three EC towers with two capturing homogeneous cotton (C3) and sugarcane (C4) fluxes, and one capturing their combined effect were used to evaluate the footprint models.

8.2 Study Area and Data Analysis

Field-scale experiments were performed in privately owned agricultural fields located at Nandi Kandi Village, Telangana, India. The annual average precipitation of the study area is 795 mm with the month July receiving the most precipitation of 170.2 mm and January receiving the least precipitation of 5.1 mm. The study area has the BSh (Köppen-Geiger climate classification) and receives an average of 48.8 days of precipitation, with the most precipitation occurring in July with 10.5 days and the least precipitation occurring in January with 0.4 days. The annual average temperature of the study area is 26.4 °C with May (32.6 °C) as the warmest and December as the coolest month (21.3 °C). Two contiguous fields comprising of cotton (C3 plant) and sugarcane (C4 plant) were selected for the experiments. The cotton field has an area of 3 acres with a canopy height of 1.0 m and was planted in June 2018. Sugarcane has an area of 2.5 acres with a canopy height of 2.2 m and was planted in January 2018. Both the crops are drip irrigated at an average frequency of 7 days. Experiments were performed using three EC flux system is composed of a 3D sonic anemometer and an open-path fast response infrared gas analyzer (IRGASON-EB-IC, Campbell Sci. Inc., USA) to measure CO_2 and H_2O fluxes. Raw data were collected with a logger (CR1000, Campbell Sci. Inc., USA) at 10 Hz frequency and averaged over half-hour intervals for use with computations. Additionally, slow response meteorological variables including precipitation (TE525-L-PTL, Tipping Bucket, Campbell Sci. Inc., USA), soil heat flux (HFP01SC-L-PT-L, Campbell Sci. Inc., USA), solar radiation (CNR 4, Campbell Sci. Inc., USA), soil moisture (CS616-L-PT-L, Campbell Sci. Inc., USA), and photosynthetically active radiation (LI-190R, LI-COR Inc., USA) were obtained at the 30-min interval. Two EC flux towers are placed at the center of C3 and C4 croplands at heights of 3 m and 4 m, respectively, and a third EC flux tower is placed at a height of 7 m to capture their combined effect (Fig. 8.1). Flux data was collected during the period 08–12 October 2018.

Primary data processing was performed on half-hour mean fluxes using Eddypro post-processing software (Version 6.2.0, LI-COR, USA). A number of standard flux corrections including (a) removal of bad data, (b) tilt corrections on sonic measurements, (c) frequency response corrections, and (d) Webb–Pearman–Leuing (WPL) corrections were applied on fast response measurements. The corrected half-hour fluxes have spiked due to unrealistic meteorological measurements, leading to the application of secondary corrections to remove spurious data and fill the gaps with reasonable estimates.

We used the REddyproc package developed in an open-source 'R' environment along with MATLAB script to perform secondary data processing that includes: (a) flux spike removal, (b) removal of negative nighttime CO_2 fluxes, (c) friction velocity correction, and (d) gap filling and uncertainty analysis. Windrose diagram (Fig. 8.2) of the study region observed during the data-period confirms that a westerly (270° from North) flowing wind predominates in the study area, and hence, all the sensors were placed accordingly, to capture the maximum possible fluxes.

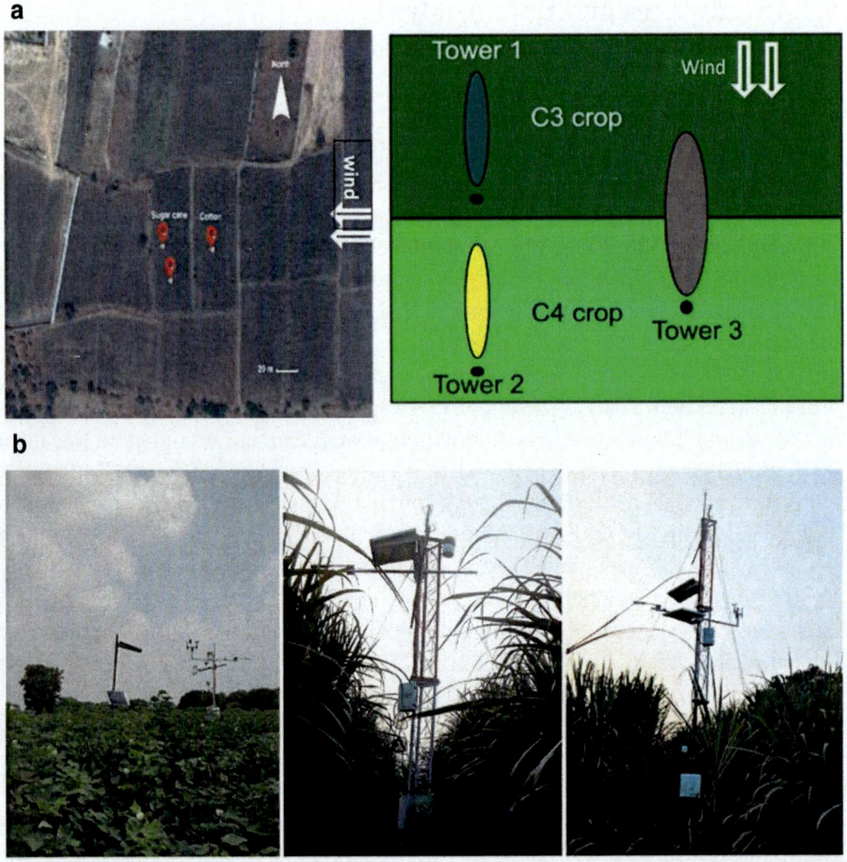

Fig. 8.1 **a** [Left] Distribution of agricultural farms within the study area, and position of three flux towers used in the analysis. Towers '1' and '2' (3 and 4 m height) capture homogeneous fluxes from cotton and sugarcane, respectively, whereas tower '3' (7 m height) captures the combined fluxes from the two fields. [Right] Typical layout and footprint from contrasting in a mixed fetch environment. **b** Eddy covariance flux tower installed in (col: 1) homogeneous cotton field (measurement height of 3 m), (col: 2) homogeneous sugarcane field (measurement height of 4 m), and (col: 3) heterogeneous sugarcane-cotton fields (measurement height of 7 m) in the Nandi-Kandi village

8.3 Methodology

The 10 Hz frequency diurnal flux data from the three EC flux systems were processed and averaged for half-hourly interval. After the quality checks (Foken et al. (1996)) only, the good data were considered for the flux footprint estimation. Flux footprints for both unstable and stable atmospheric conditions were estimated using three analytical models (Schuepp model, Korman and Meixner model, and Hsieh model) and the experimental data collected. The flux footprint predicted by the analytical models were then compared with the results of the flux footprint estimated

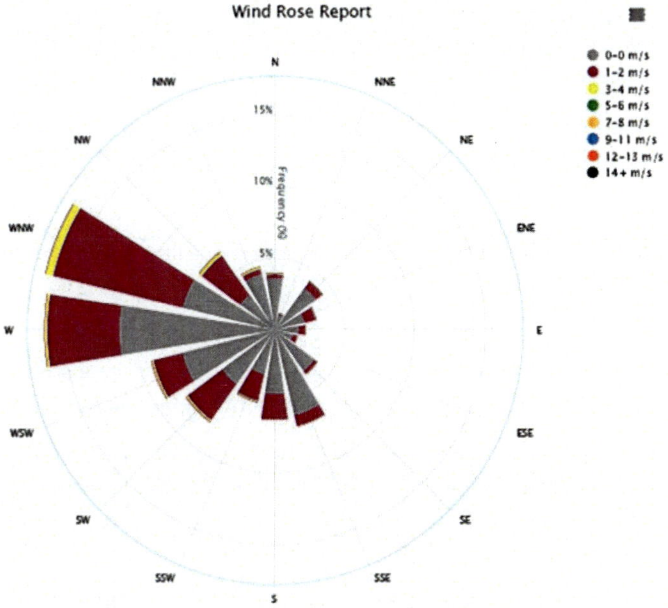

Fig. 8.2 Windrose diagram showing predominant wind direction within the study area during the monitoring period

by Lagrangian particle distribution model. The models used to predict flux footprint include:

(1) *Lagrangian (Kljun) model* considers release of a point source, and tracts the trajectories downwind of the source toward the measurement location, with footprint function given by

$$\hat{F}^{y*} = a\left(\hat{X}^{*} - d\right)^{b} \exp\left(\frac{-c}{\hat{X}^{*} - d}\right)$$

where *a*, *b*, *c*, and dare fitting parameters, $\widehat{F}^{y*}$ is the crosswind-integrated footprint function, and X^{*} is along with wind distance.

(2) *Schuepp model* estimates weighing factors of footprint function for scalar flux and concentration as an analytical solution of advection–diffusion equation, given by

$$f(x) = \frac{U z_m}{ku * x^2} \exp\left[\frac{U z_m}{ku * x}\right]$$

where U is mean horizontal wind speed, z_m is measuring height, u^* is frictional velocity, k is von Karman constant, and x is the horizontal upwind distance of sources from the measurement system.

(3) *Korman and Meixner model* considers homogeneous and stationary flow conditions over homogeneous terrain, given by

$$f(x) = \frac{1}{\Gamma(\mu)} \frac{\xi^{\mu}}{x^{1+\mu}} e^{\left(\frac{\xi}{x}\right)}$$

where Γ is the Gamma function, μ is a parameter that depends on atmospheric stability conditions and ξ is a coordinate incorporating measurement height.

(4) *Hsieh model,* an approximate analytical solution to advection–dispersion equation, largely suited to thermally stratified flows, given by

$$f(x, zm) = \frac{1}{k^2 * x^2} D Z u^P |L|^{1-P} \exp\left(\frac{-1}{k^2 * x} D Z u^P |L|^{1-P}\right)$$

$$L = -\frac{u^{*3}}{k \frac{gQ}{T\rho cp}}$$

where $k = 0.4$ is the von Karman constant, u^* is friction velocity, Q is turbulent heat flux, T is surface temperature, c_p heat capacity, and z is the measuring height.

Atmospheric stability (ζ) is defined as:

$$\zeta = \frac{Z - d}{L}$$

where Z is sensor height above the ground, D is zero plane displacement height, and L = Obukhov length (Table 8.1).

Table 8.1 Classification of stability of atmosphere based on calculated atmospheric stability (Z-d/L). Atmospheric stability condition defines the size of the flux footprint

Category	(Z – d/L)
Extremely unstable	≤1
Unstable	−1 to −0.05
Neutral	−0.05 to 0.05
Stable	0.05–1
Extremely stable	>1

8.4 Results and Discussion

Table 8.2 shows detailed meteorological and surface roughness conditions measured by eddy covariance stations during stable and unstable atmospheric conditions. The partition of fluxes were done using the parameter ($Z - d/L$), with a value <0 representing unstable (Daytime), and >0 representing stable (Nighttime) conditions. Figure 8.3 shows the variation in the meteorological parameters during the experiment period. Frictional velocity (u*) varies with the surface roughness, which varies with crop type and can be observed from Table 8.2.

Figure 8.4 represents the crosswind-integrated footprint for CO_2 fluxes obtained from three different analytical flux footprint models (Hsieh, K & M and Schuepp) and Lagrangian particle distribution Kljun (LPDM-B) flux footprint model for two different atmospheric stability condition (unstable and stable) in three different canopies) C3 cotton cropland, (b) C4 sugarcane cropland, and (v) C3–C4 mixed cropland. It can be observed that the CO_2 flux footprint or fetch varies with the change in atmospheric stability condition and approximately 75% of the flux source is confined within a range of 30–40 times the effective measurement height (Zm – d). The flux footprint in the stable atmospheric condition is too large with respect to the unstable atmospheric condition as in the stable condition, i.e., during a clear and calm night, the air within a stable layer is not turbulent and in unstable condition parcel of air are warmer than the surrounding air and together with turbulent air enhances the vertical motion of the CO_2 eddies.

From Fig. 8.4, it can be observed that Hsieh and K&M model are underestimating the upwind peak flux footprint distance (X_{max}) and the magnitude of peak flux (f_{max}). Schuepp model is underestimating f_{max} with respect to backward Lagrangian particle distribution model (Kljun (LPDM-B) for C3 and C4 crops during unstable conditions.

K&M model was observed to overestimate X_{max}, as the model neglects along with wind turbulent diffusion. Our results conclude that ignoring the along-wind turbulent diffusion can lead to an underestimation of flux contribution close to flux tower (Rannik et al. 2000) and increases the extent of flux footprint fetch (Schmid 2002). The performance of analytical models is evaluated using three residual statistical parameters including RMSE, NSE, and correlation coefficient R^2. The model that results in lowest RMSE, highest NSE, and R^2 was chosen to accurately represent the flux footprint. Schuepp model performs best to the Lagrangian model in both C3 and C4 crop field in unstable condition and Hsieh model performs best in mixed fetch condition (C3–C4) by considering the heterogeneity in the flux source in both stable and unstable atmospheric stability condition. Hsieh model also performed best for the homogenous cotton crop (C3) and gives a good estimate of flux footprint in stable condition (Table 8.3).

Table 8.2 List of meteorological variables mean wind speed (u [ms^{-1}]), frictional velocity (u^* [ms^{-1}]), Obukhov length (L[m]), roughness length (zo[m]), displacement height (d[m]), atmospheric stability ($Z - d/L$), evapotranspiration (ET [mm/hr]), and vapor pressure deficit (VPD [hPa]) measured by eddy covariance station (Zm = 3, 4, and 7 m) considering unstable (day time) and stable (night time) atmospheric conditions at Nandi Kandi flux site

Day time									
Crop	DOY	u (ms^{-1})	u^* (ms^{-1})	L(m)	z0(m)	d(m)	($Z - d/L$)	ET	VPD
Cotton (C3)	282.000	1.597	0.344	−154.149	0.220	1.020	−0.019	0.058	6993.525
Sugarcane (C3–C4)	282.000	0.430	0.083	−0.982	0.500	3.000	−0.489	−0.112	2443.264
Mixed fetch (C3–C4)	282.000	0.957	0.032	−0.630	0.600	2.680	−3.682	−0.004	2643.263
Night time									
Crop	DOY	u (ms^{-1})	u^* (ms^{-1})	L (m)	z0(m)	d(m)	($Z - d/L$)	ET	VPD
Cotton (C3)	282.000	1.289	0.343	78.845	0.220	1.020	0.038	0.024	5595.206
Sugarcane (C3–C4)	282.000	0.716	0.024	0.275	0.500	3.000	1.744	0.009	1784.401
Mixed fetch (C3–C4)	282.000	0.733	0.012	0.159	0.600	2.680	14.572	0.003	2197.845

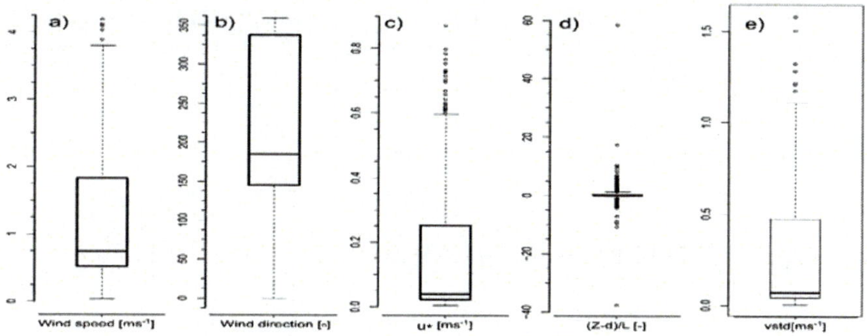

Fig. 8.3 Boxplots of **a** wind speed, **b** wind direction, **c** friction velocity $u*$, **d** atmospheric stability parameter $(Z - d/L)$, and **e** the standard deviation of the lateral wind component during October 08–12, 2018

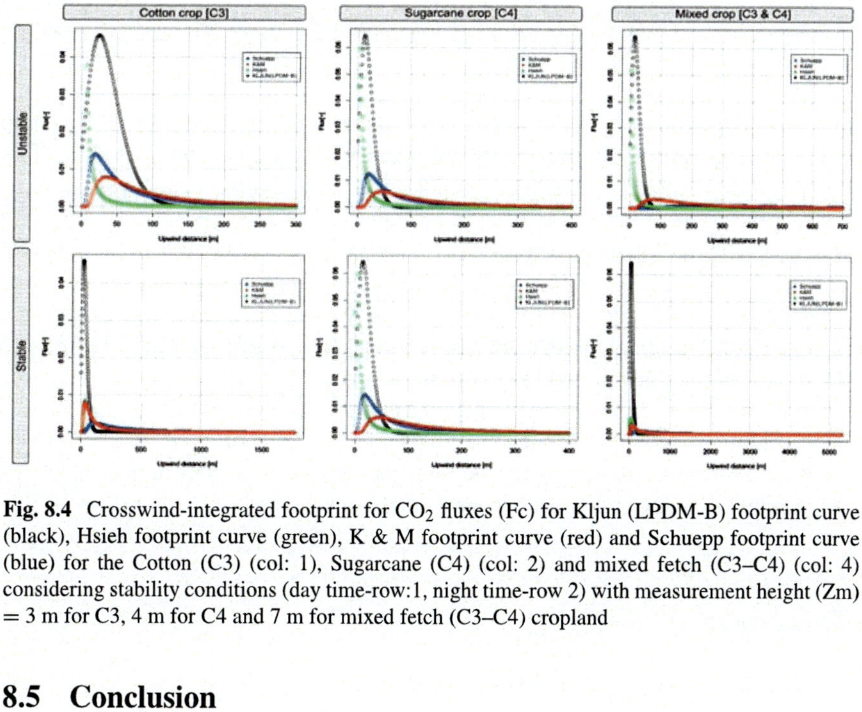

Fig. 8.4 Crosswind-integrated footprint for CO_2 fluxes (Fc) for Kljun (LPDM-B) footprint curve (black), Hsieh footprint curve (green), K & M footprint curve (red) and Schuepp footprint curve (blue) for the Cotton (C3) (col: 1), Sugarcane (C4) (col: 2) and mixed fetch (C3–C4) (col: 4) considering stability conditions (day time-row:1, night time-row 2) with measurement height (Zm) = 3 m for C3, 4 m for C4 and 7 m for mixed fetch (C3–C4) cropland

8.5 Conclusion

Flux measurement and footprint modeling in the heterogeneous canopy with variation in atmospheric stability and surface roughness conditions is challenging. In this study, three analytical flux footprint models are compared with the Lagrangian flux footprint model applicable to typical fragmented croplands of the Indian setting.

Table 8.3 Statistics of flux footprint model performance considering residual statistical parameters for C3 at (Zm = 3 m), C4 (Zm = 4 m), and mixed cropland (C3–C4) (Zm = 7 m) in different atmospheric stability conditions

Model comparison w.r.t. KLJUN(LPDM-B)							
Analytical Model	RMSE	NSE	Correlation	RMSE	NSE	Correlation	
	Cotton(C3): Day time (Unstable condition)			Cotton(C3): Nighttime (Stable condition)			
Hsieh	0.0126	0.1648	0.2528	0.0053	0.2914	0.7649	
K&M	0.0084	0.0521	0.7824	0.0054	0.2895	0.7400	
Schuepp	0.0073	0.0951	0.9541	0.0063	0.1929	-0.0103	
	Sugarcane(C4): Day time (Unstable condition)			Sugarcane(C4): Nighttime (Stable condition)			
Hsieh	0.0119	0.6244	0.5759	0.0125	0.5351	0.5510	
K&M	0.0140	0.4261	0.0497	0.0143	0.3246	0.0925	
Schuepp	0.0121	0.5334	0.6692	0.0119	0.4530	0.7992	
	Mixed fetch(C3-C4): Day time (Unstable condition)			Mixed fetch(C3-C4): Nighttime (Stable condition)			
Hsieh	0.0247	0.4299	0.3530	0.0060	-2.2641	0.8233	Good
K&M	0.0239	-0.3994	-0.9333	0.0063	-2.2837	0.5200	Average
Schuepp	0.0240	-1.0154	-0.5678	0.0066	-2.3682	-0.0157	Bad

We have experimented in C3 (cotton), C4 (sugarcane), and mixed (C3–C4) croplands having different homogeneity and surface roughness conditions. Flux footprint estimated using Lagrangian particle distribution method that considers the heterogeneity of fetch was used as a reference. To select the suitable model for given agro-meteorological conditions, the performance of three analytical models was evaluated with respect to the Lagrangian model. All analytical models were observed to be sensitive to surface roughness, atmospheric stability, and measurement height. We conclude that the conventional thumb rule of fetch measurement (1:100) is not applicable to the study area as about 75% of the fluxes comes from source, i.e., 30–40 times the effective measurement height. For homogenous cropland condition, Schuepp model can be used in unstable condition and Hsieh model in stable condition. However, for heterogeneous condition, Hsieh model can be used in both unstable and stable atmospheric conditions.

References

Arriga N, Rannik Ü, Aubinet M, Carrara A, Vesala T, Papale D (2017) Experimental validation of footprint models for eddy covariance CO_2 flux measurements above grassland by means of natural and artificial tracers. Agric For Meteorol 15(242):75–84

Aubinet M, Grelle A, Ibrom A, Rannik Ü, Moncrieff J, Foken T, Kowalski AS, Martin PH, Berbigier P, Bernhofer C, Clement R (1999) Estimates of the annual net carbon and water exchange of forests: the EUROFLUX methodology. In: Advances in ecological research, vol 30, pp 113–175. Academic Press

Finnigan JJ (2004) A re-evaluation of long-term flux measurement techniques part II: coordinate systems. Bound-Layer Meteorol 113(1):1–41

Foken T, Leclerc MY (2004) Methods and limitations in validation of footprint models. Agric For Meteol 127(3–4):223–34
Foken T, Wichura B (1996) Tools for quality assessment of surface-based flux measurements. Agric for Meteorol 78(1–2):83–105
Horst TW, Weil J (1992) Footprint estimation for scalar flux measurements in the atmospheric surface layer. Bound-Layer Meteorol 59(3):279–296
Hsieh CI, Katul G, Chi TW (2000) An approximate analytical model for footprint estimation of scalar fluxes in thermally stratified atmospheric flows. Adv Water Resour 23(7):765–772
Kaimal JC, Finnigan JJ (1994) Atmospheric boundary layer flows: their structure and measurement. Oxford university press
Kljun N, Calanca P, Rotach MW, Schmid HP (2004) A simple parameterisation for flux footprint predictions. Bound-Layer Meteorol 112(3):503–523
Kolari P, Pumpanen J, Rannik Ü, Ilvesniemi H, Hari P, Berninger F (2004) Carbon balance of different aged Scots pine forests in Southern Finland. Glob Change Biol 10(7):1106–1119
Pasquill F, Smith FB (1983) Atmospheric diffusion: study of the dispersion of windborne material from industrial and other sources. Wiley, New York
Rannik Ü, Aubinet M, Kurbanmuradov O, Sabelfeld KK, Markkanen T, Vesala T (2000) Footprint analysis for measurements over a heterogeneous forest. Bound-Layer Meteorol 97(1):137–166
Schmid HP (1994) Source areas for scalars and scalar fluxes. Bound-Layer Meteorol 67(3):293–318
Schmid HP (2002) Footprint modeling for vegetation atmosphere exchange studies: a review and perspective. Agric For Meteorol 113(1):159–183
Van de Boer A, Moene AF, Schüttemeyer D, Graf A (2013) Sensitivity and uncertainty of analytical footprint models according to a combined natural tracer and ensemble approach. Agric For Meteorol 15(169):1–1
Wilson JD, Swaters GE (1991) The source area influencing a measurement in the planetary boundary layer: the "footprint" and the "distribution of contact distance." Bound-Layer Meteorol 55(1–2):25–46

Chapter 9
Water Resources Assessment Issues and Application of Isotope Hydrology in North East India

Prem Ranjan, Pankaj Kumar Pandey, Vanita Pandey, and Pema Tshering Lepcha

Abstract The freshwater resources are likely to be severely impacted by climate change as reported by many researchers around the globe. India already facing a shortage of freshwater resources due to rapid urbanization and industrialization and acceleration of economic development activities. Assessment of water resources plays a key role in the development of the economy of any nation. Northeastern states of India are blessed with huge water resources. The region faces inconsistent circumstance as devastating floods in monsoon months and water scarcity in non-rainy periods at many places. Most of the areas having high reliefs or undulating terrains where rainwater does not get sufficient time to infiltrate into the soil. As a result, quick runoff is dominant in the region, and ultimately reduces the recharge of springshed, and results in the reduction of discharge of many springs and streams during the non-rainy season. To overcome freshwater stress in the region, a proper understanding of the water cycle, its management, and development of various infrastructure is highly required. Though some problems could not be solved due to the lack of hydrological data like spring discharge data, water quality, etc. The alternate use of environmental isotope technologies aids researchers to estimate origin, recharge source age, and its movement within the hydrologic cycle can be of ultimate solutions. The stable isotopes (Deuterium and Oxygen-18) are excellent indicators of the water circulation, whereas radioisotopes (Tritium and Carbon-14) distinct value in detecting the mean residence time (MRT). This methodology has special value in terms of its cost-effectiveness and the investigative encouragement of the specialists. This paper presents the use of isotopic hydrology for sustainable development of the water resources of northeastern region of India.

Keywords Springshed · Environmental isotopes · Groundwater · Spring water · Hydrogen and oxygen

---

P. Ranjan (✉) · P. K. Pandey · V. Pandey · P. T. Lepcha
Department of Agricultural Engineering, North Eastern Regional Institute of Science and Technology, Arunachal Pradesh, Nirjuli 791109, India

R. Jha et al. (eds.), *Water Resources Management and Reservoir Operation*, Water Science and Technology Library 107,
https://doi.org/10.1007/978-3-030-79400-2_9

9.1 Introduction

In the recent few years, every part of India is facing the problems of hydrological extremes such as floods and droughts frequently. More than 100th districts of the country are still witnessing a drought like-situation, due to failure of monsoon rainfall (Jitendra 2015). According to the Indian Meteorological Department, a mild to extremely dry conditions in 404 Indian districts have witnessed due to poor rainfall in past few years. It is alarming the coming water crises in the country. Many regions of India facing water crisis in the peak of summer. The Himalayas are known as the "water towers of the Earth." The Himalayan range provides water to the millions of people at the downstream end through the perennial rivers (Vashisht 2015). It is reported that in the Himalayan region rainfall will be slightly increased along with greater intensity (Rees and Collins 2006; Singh et al. 2008; Tambe et al. 2012; Agarwal et al. 2012). In the past five decades, it has been witnessed that the rainfall pattern has changed from low-intensity longer-duration events to high-intensity smaller-duration events (Vashisht 2017; Vashisht and Ranjan 2020). Kusre et al. (2017) reported that in Sikkim more than 90% average annual rainfall is concentrated during April to September. On the other hand, they have witnessed that less than 2–8 mm rainfall occurring at some stations in the non-rainy period. On the monthly basis, 70–85% of the rainfall is reported from April to September, whereas the remaining 15–30% of the rainfall is observed from October to March. Apart from rainfall, change in land use and land cover is another challenge in water resource management. Recharging capacity of catchments is reducing due to deforestation, cutting of hills for new construction (such as a house, roads, and mining), cleaning of the forest for jhoom cultivation, forest fires, overgrazing and trampling by livestock. So, one of the main reason behind the reduced in groundwater recharge is the change in land use and land cover that results in higher runoff (Negi and Joshi 2002).

Nowadays, new methodologies and techniques have been developed in the field of water resource management sector. In the context of computer-aided research, remote sensing technique with GIS can be incorporated for any water-related studies. This can reduce the complexity involved in studying water resources management studies at any level. Isotope hydrology and hydrogeology is an interdisciplinary science and emerging discipline with expanding investigation tools for many environmental problems (Barbieri 2019). It is an alternate solution to many problems and its importance felt more in recent years. The water sector has facing different problems such as depleting groundwater, deterioration in water quality, and many other erratic natural events which affect the hydrological cycle. The traditional standard methods often fail to provide proper information to many of these problems, whereas use of isotope method provides a clear picture and help in finding a suitable solution (Barbieri 2019; Kumar 2013). It is now possible, by using of isotopes as conservative tracers of groundwater flow paths, its sources of regions, resident time in the watershed, identifying the recharge areas, chemical reactions in surfaces and another environmental situation. Use of isotopes in hydrology is based on the general concept of "tracing" in which either manually injected or naturally occurring isotopes are utilized. Environmental isotopes referred to the naturally occurring elements in waters contain about seventeen hundred stable and unstable or radioactive isotopes. Environmental isotopes either stable or unstable (radioactive) have a distinct advantage over artificially tracers in the sense they encourage the investigation of different hydrological measures on a larger temporal and spatial scale through their characteristic circulation in a hydrological system. Hence, environmental isotopes technique is unique in regional studies to find time and space integrated attributes, while artificial tracers are effective for site-specific and local applications. The most habitually utilized environmental isotopes contain deuterium (^{2}H), tritium (^{3}H), oxygen (^{18}O), and carbon (^{13}C, ^{14}C) occurring in water as constituents of dissolving organic and inorganic compounds (Clark and Fritz 2013). Hydrogen and oxygen (stable isotopes), being building blocks of the water molecule and hydrological cycle, show a characteristic isotope signature in water that can be procured as tracers of groundwater recharge (Price and Swart 2006; Lee et al. 1999; Clark and Fritz 2013) and its signature is not modified by other biosphere factors like rock–water interaction (Sidle 1998). In mountainous areas, $\delta^{18}O$ and $\delta^{2}H$ are used to distinguish precipitation derived groundwater as a result of altitude effect. For determining the recharge area of the groundwater (Fontes et al. 1967; Stahl et al. 1974), the origin of springs (Dhakal et al. 2014), delineation of spring catchment, to estimate the recharge from precipitation groundwater flow from higher slopes (Jeelani and Shah 2014).

9.2 Water Resources Assessment Issues in Northeastern Hilly States

Northeastern states of India face water scarcity despite receiving intensive orographic rainfall, because of both the physical and anthropogenic reasons. People's general perception has water resource management practices that are relevant to water scares region, but this a wrong opinion need of these practices which has been also felt in high precipitation regions. Cherrapunji has the highest rainfall in the world, but this place suffers from drought for 9 months of the year (Agarwal 2000; Kumar et al. 2005). Most of the hilly states of the northeastern part of India has high rainfall likewise Meghalaya, but in the non-monsoon period, they felt water-scarce situation (Kusre et al. 2018). Highly undulating land is one of the main reasons for the shortage of water in these hilly states. These hilly states suffer from the shortage of water due to highly undulating terrain, generally in the non-monsoon period when there is little rain or no rain (Lepcha et al. 2021). Also, the other main cause of the declining of springs discharge is steep terrain of these hilly state. In Sikkim, the elevation range is shown in Fig. 9.2 (DEM of Sikkim) which indicated that steep terrain, it is also the main cause of less opportunity of recharge of springs and water conservation practice is tough tasks, this scenario exists in all states of Northeast India. Fig. 9.3 shown the DEM of Meghalaya. The water-scarce situation is causes of forest fires, drop in productivity, incidences of pest and diseases, and shortage of drinking water; it directly affects the economy of the states. Two springs located in Lekhi village and Barapani of Papumpare district of Arunachal Pradesh is shown in Fig. 9.1.

Springs (shown in Fig. 9.1) are the main source water, most of the communities are dependent on this important source (NITI Ayog 2017). It is an essential source for the existence of human beings in this region. Apart from human and domestic animals, it has also important for wildlife, ecosystem (Tambe et al. 2012). They are not just part of the heritage, but the health of the rivers also depends on the health of the springs. It formed when geologic, hydrologic, or living forces cut the underground layers of

(a) (b)

Fig. 9.1 Springs located at **a** Lekhi village, **b** Barapani, Papumpare district, Arunachal Pradesh

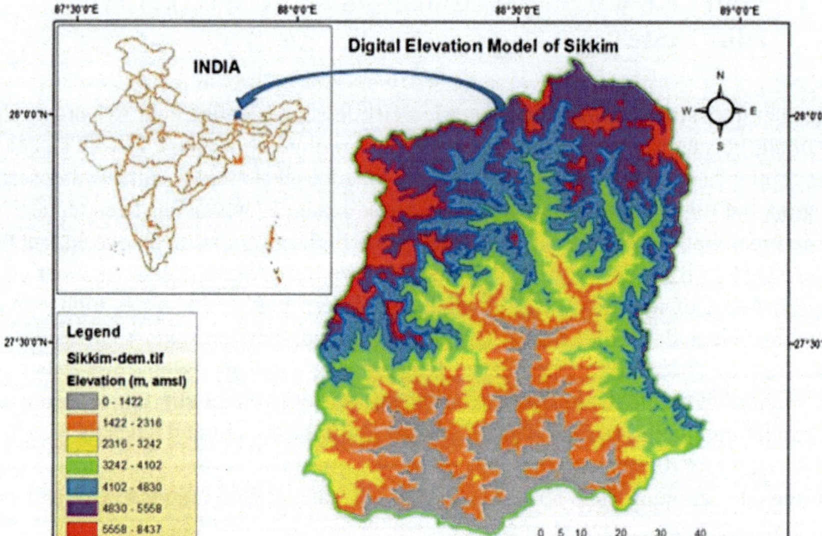

Fig. 9.2 Digital elevation model of Sikkim

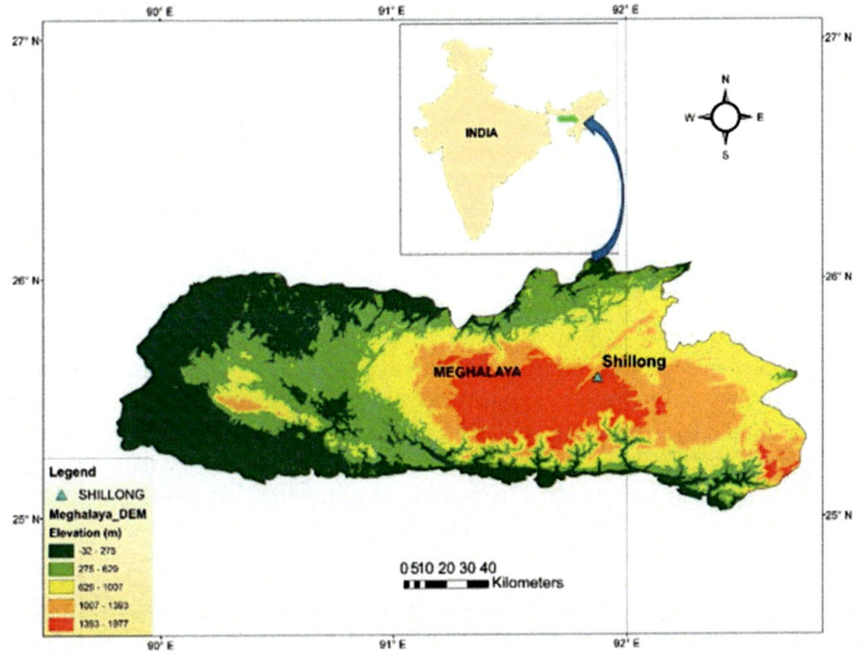

Fig. 9.3 Digital elevation model of Meghalaya

soil and rock where water is in movement (Giberson et al. 2013). Each spring is not quite the same due to the slope and geological structure of the springshed such as its type, recharge area, and discharge rate. Based on these characteristics, spring has different classifications such as depression, contact, seepage, gravity, and fracture or fault spring (Mahamuni and Upasani 2011).

A questionnaire survey is conducted in Sikkim by Kusre et al. (2017), and they observed that most of the people have felt water scarcity situation in the non-monsoon period. Ninety percent of respondents expressed that from December to March, water availability is very less. Also, it has been reported that the discharge of springs is declining or drying in the December to March period (Tambe et al. 2012). The main reason of declining of springs are climatic vulnerability (especially rainfall), change in land use/land cover, socioeconomic and demographic changes (Tambe et al. 2012; Ranjan and Pandey 2020). In some areas, where villages are situated at hilltop, the problem has become so acute that peoples are leaving or moving their villages.

It was noticed by local natives that from the last four to five decades, the flow rate of spring and streams is sharply declined. Now the situation is so worst that the springs, lakes are dry in non-rainy period, and also perennial rivers' flow rate is diminished. To sustain in this water-scarce situation, they have no other choice. To collect water, peoples travel long distances (Vashisht 2015). In rural areas, women and children wake up early morning to travel long distances to collect water to fulfil the daily household requirement. According to a report, 72% women and 14% of children have to bear the responsibility of collecting potable water. About, 60% of the women have to walk half km, and 10% of them have to walk 4 km o fetch water (Vashisht and Sharma 2007). In this water-scarce scenario, farmers are leaving abandon the agricultural land due to an insufficient quantity of water for irrigation.

9.3 Springshed Development and Challenges

To survive in this water-scarce scenario, there is a need to adopt different water management practices. Such as the installation of water harvesting structure, rejuvenation of springs, recharging of groundwater, transferring of water from excess to shortage area, and spring development practices (Vashisht 2015; Agrawal et al. 2015; Tambe et al. 2013). Implementing of springshed development programs has many challenges such as inventory of spring in remote areas, accurately identification of recharge areas, in-situ rainwater harvesting in agricultural land, developing the local capacity building, and public financing. The major goals of springshed development are

- Stop deforestation, planting trees, increase vegetation cover and soil organic matter, reduce soil erosion, and reduce runoff velocity,
- In-situ rainwater harvesting by bunds, trenches, ponds, and check dams,
- Increase infiltration opportunity time to naturally recharge the aquifer,
- Protection of springshed to allow ecological restoration and prevent contamination of the groundwater and runoff water,
- Develop suitable water storage structure,
- Aware and train the community about springshed development and management practices.

Information of reliability of springs is essential to adopt any water management practices. There are various methods, models, and indices which are available to determine the reliability of springs such as (i) Time-lag from the incidence of particular rainfall to consequent change in the spring discharge, (ii) the amount of rainwater stored in the aquifer; (iii) maximum to minimum discharge ratio of spring in a year; and (iii) prediction of spring discharge rate and yield for upcoming recession period. So, to assess the spring reliability rainfall, the discharge rate is required. To quantify the spring reliability based on historical data, different mathematical techniques are available (Vashisht 2015, 2017).

The discharge rate of any springs depends on several factors, such as natural and man-made. But the main factor is the amount of water stored in an aquifer, which depends on rainfall occurrence in spring catchment and locally recharged (Ranjan and Pandey 2020). Spring will eventually dry up, if there is no rainfall or recharge in spring catchment or if the recharge rate is lesser than pumping rate. About 90% of the total rainfall occurred from June to October or April to September, in this duration aquifer is recharged. Precipitation in the non-monsoon period is negligible or sometimes it completely absent. So, recharge non-rainy season is negligible or if some amount of rainfall occurs in this duration it may evaporate. The best condition of recharging of springs is when temperature and evaporation are low and the ground is more absorbent.

9.3.1 In-Situ Spring Water/Rainwater Harvesting

Springs are drying due to less recharge or no recharge of the aquifer. The major reason for not recharging or replenishment of aquifers are deforestations, change in precipitation pattern, land cover, and soil property. Due to deforestation and change in precipitation (such as high density and short duration), rainwater is not getting sufficient time for infiltration or deep percolation, only a few quantities of water infiltrate, and most of the flushes in valleys and streams. The high quantity of runoff is one of the major reason for landslides.

Construction of trenches, pit and check-dam, plantation of local tree species, minimizing grazing, cutting of fuelwoods and grasses and installing the fencing around the springshed are the common measures for development of springs and groundwater. Developing trenches (shown in Fig. 9.4a), CCT, pits, and check dams (shown in Fig. 9.4b) is generally used in rejuvenation programs. These structures break the runoff velocity and increase the infiltration opportunity time to recharge the aquifer. Apart from the rejuvenation program, harvest available water when it is excess and construct storage structures. Apart from these techniques, local communities of the Himalayan region has also developed many in-situ water harvesting practices, depending upon location specifics and named them accordingly.

(a) (b)

Fig. 9.4 **a** Trenches excavated at hill slope, and **b** Bamboo check dam constructed in stream

9.3.2 Springshed Protection

Protect the springshed catchment area, grow preferred natural vegetation, and disregard the human activities like agricultural uses which increase the chances to the inflow of pesticides and erosion risks. Adopt different traditional and modern techniques to protect the recharge area. Such as install fences surrounding of recharge area to avoid man and animal interference, avoid any garbage dumping sites, wastewater discharge in springshed, avoid plant and hill cutting for new construction (roads, hotels, picnic spots, and industries) in springshed, avoid pesticides and insecticides in agricultural filed, plant local, and less water consumption trees. Local communities and water user groups are also using traditional knowledge to protect the spring. Communities of this reason assume water sources are holy places. They worship these sources as a goddess. These practices are also helpful to protect spring from contamination because people think this is a holy place. So, they not throw any litters and garbage at that site. But these practices are limited at the only source point.

9.3.3 Major Challenges, Conflicts and People Participation in Spring Development and Management Practices in Northeast Himalayas Region

In Himalayan region, travelling and fieldwork are a very hectic task due to the uneven topographic. Without proper climbing equipment and local guide, many places are inaccessible due to high relief. The complex geology and repeated structural disturbances and thick forest coves make it more difficult to develop a geological map and its study. Also, disasters such as landslides and floods add more difficulties in research work. Unavailability of spring data is one of the major challenges of development practices. Lack of historical data of springs such as location, discharge rate, types of spring availability, water quality, and other aquifer properties. So, there is no database for any planning, development, and management practices. The major cause of the lack of historical data is the lack of awareness about this important source. Most of the communities living in this region are depended on these important sources, but data collection is never a priority. Communities are not aware of the upcoming water-scare situation. So, these are the major reason for such data gaps. As per the geohydrology of the area, the recharge area of springs may fall in the forest, in administrative boundaries of beneficiary communities or maybe in any other villages or maybe in any other private land. Apart from geohydrology land, use of recharge area is also a major challenge, and it may be agricultural and horticultural land, reserved forest, private forest, and wastelands. So, this is one of the causes of conflicts between two villages or communities or landowner. In the case of government property or forest land, there is a need for permission from the concerned department or forest department for any development practices. The methodology for executing the revive measures cannot be uniform in all the cases referenced, and hence, it poses challenges for the implementing organizations.

9.4 Isotope Hydrology Applications in Water Resource Management

Investigation of groundwater from conventional techniques fail to provide proper information such as recharge sources, mechanism, and deposition. Adopting the new technologies such as isotope hydrology helps to investigate the clear picture or proper information of groundwater flow paths, its sources of regions, resident time in the watershed, identifying the recharge areas, chemical reactions in surfaces, and another environmental situation. Environmental isotopes, both stable and unstable (radioactive) isotopes, are used in water resources assessment. Stable isotopes usually measured by Isotope Ratio Mass Spectrometer (IRMS), in terms of the isotope ratios of the less abundant to more abundant. Radioactive (or unstable) isotopes are measured by counting the number of atoms (using accelerator mass spectrometer) or counting their radioactive decays (by liquid scintillation counter) in a given sample. It is an effective technique to obtain the necessary critical hydrological information such as (1) Determination of age, velocity, and path of flow; (2) Origin or source of recharge; (3) Possible interconnections between different aquifers, and surface water and groundwaters; and (5) Aquifer properties such as porosity, transmissivity, and dispersivity. The isotopic concentrations are expressed as the variations between the measured ratios and the standard sample. The isotopic compositions of water are denoted by deviations (δ) and expressed in per mil (‰) from a reference standard mean ocean water (SMOW) as shown below:

$$\delta(^0/_{00}) = \left\{ \frac{\left(\frac{\text{Abundnce of less (heavier) isotope}}{\text{Abundance more abundant (lighter) isotope}}\right)_{\text{Sample}}}{\left(\frac{\text{Abundance of less abundant (heavier) isotope}}{\text{Abundance of more abundant (lighter) isotope}}\right)_{\text{Standard}}} - 1 \right\} \times 1000$$

or

$$\delta(^0/_{00}) = \left(\frac{R_{\text{sample}}}{R_{\text{standard}}} - 1 \right) \times 1000 \quad (9.1)$$

As the defined δ values are small numbers, they are expressed in ‰ (per mill).

9.4.1 *Global Meteoric Water Line (GMWL) and Local Meteoric Water Line (LMWL)*

GMWL Eq. (9.2) is recognized by Craig (1961) based on global relationship between δD and δ^{18}O in natural meteoric water. Water samples of rainfall, snow, and river are collected from all over the world and analyzed for their isotopic composition. Based on Eq. (9.2), deviations of isotopic composition from equilibrium process to local or regional line can be defined. The meteoric water line may vary with a place to place depending on weather condition. Further, Rozanski et al. (1993) have developed Eq. (9.3) based on Eq. (9.2) by incorporating the monthly average values from different stations belonging to Global Network of Isotopes in Precipitation monitored by World Meteorological Organization and Atomic Energy Agency.

$$\delta D = 8 \times {}^{18}O + 10 \tag{9.2}$$

$$\delta D = 8.1\,(\pm 0.07) \times \delta 18O + 11.27\,(\pm 0.65) \tag{9.3}$$

The GMWL represents the variation of δ^2H or δD to $\delta^{18}O$ in precipitation and can be utilized to identify the source of water, mixing, and other hydrological measures. Almost, precipitation in every part of the world follows this relationship. But in some specific areas which have different temperature and humidity or unique climate, create their special LMWL with a different slope and intercept. The slope and intercept of LMWLs depend on hydrological parameters.

9.4.2 Identify the Springs Origin and Their Recharge Area, Altitude Effect, Mean Residence Time

Springs originating naturally from unconfined aquifers were several factors such as changes in climatic conditions, land use, and demography. are affecting the discharge of springs. Communities of the northeastern hilly region are fully depending on springs. Springs are also responsible for ecological balance and baseflow of flowing rivers. The variation in the $\delta^{18}O$, δD (or δ^2H) isotopic signatures of catchment waters (streams, springs, precipitation, and snow) have been used to identify the hydrological source areas under different flow conditions based on a spatio-temporal variation of stable water isotopes which is pre-requisite for understanding the identification of recharge problem.

Dhakal et al. (2014) conducted an isotope-based study in Sikkim to identify the origin and recharge area to understand the spring's dynamics. They found that the initial results reveal that the spring discharges are strongly dependent on the rainfall pattern. The stable isotope studies of the spring samples show that the spring samples fall above the global meteoric water line as shown (Fig. 9.4) by Himalayan aquifers and evaporation enrichment during the recharging process. Electrical conductivity analysis shows that the aquifers are recharged on a local level. All spring samples fall above the Global Meteoric Water which is generally observed in the Himalayan region because rainwater is depleted. Linear regression line equation for samples collected in March 2013 is $\delta^2H = 5.5\delta^{18}O–5.9$ ($r = 0.82$, $n = 12$) which indicates evaporative enrichment of stable isotopes during the recharge process. The highest enrichment has been found in the spring named Lower Changey Source. Linear regression line equation for samples collected in May 2013 is $\delta^2H = 4.1\ \delta^{18}O–16$ which indicates more enrichment than March samples with a linear regression line of $\delta^2H = 5.5.\delta^{18}O–5.9$. The study shows that the springs are local origin and the possible recharge zone of the aquifers is within the local geographical area. They also show that the rainfall-spring interdependence which clears that the source of recharge is rainwater (Fig. 9.5).

There are several studies on the identification of origin and recharge area delineation in different Himalayan regions. In western Himalya, Jeelani (2008) observed an inverse relationship between spring and stream discharge with precipitation in Anantnag District of Jammu & Kashmir. He also observed that springs of Anantnag district are recharged by glaciers and are under threat due to the reduction in glacial coverage. Jeelani et al. (2010), Bhat and Jeelani (2015) delineated the origin of streams and springs of Liddar and Bring Bringi catchment of Kashmir Himalayas using $\delta^{18}O$. They also analyzed the effect of altitude on springs and observed multiple sources during warm months and common source during colder months. Shah et al. (2017) estimated mean residence time (MRT) using tritium method, sine wave model, and artificial tracers and reported that cold karst springs are fed with recent recharge and on the other hand, warm karst springs with a mixture of old and recent recharge. A comparison of MRT has been made between tritium and the sine wave model, longer MRT was observed in warm karst springs and shorter in cold springs. They reported that short travel time of karst spring in the region indicates the water quality deterioration and variation in the magnitude of flux. So, identified catchment area must be protected, preserved, and managed from any anthropogenic activities.

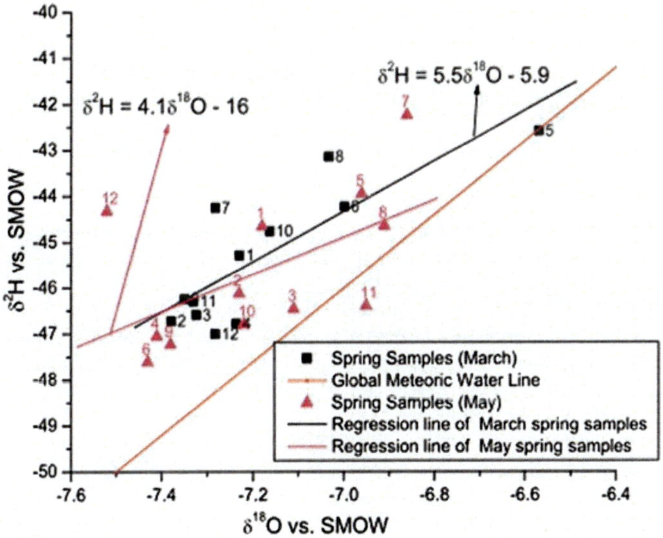

Fig. 9.5 $\delta^{18}O$ *versus* δ^2H plot of spring water samples collected in March and May 2013. *Source* Dhakal et al. (2014)

9.5 Conclusion

The northeastern hilly states communities are mostly dependent on springs and streams for freshwater. It is reported that the springs are drying up due to the uncertainty of rainfall and land use/land cover change. Water harvesting is also a tough challenge due to the steep topographic profile exist in the NE region of India. Water scarcity is increased in the winter season. At some places, women and children travel long distances to get water. To sustain in this water-scarce situation, rejuvenate or recharge springs/groundwater and protect this important source. Use traditional and modern techniques to harvest water. Now, it needs to the management of springs using new scientific methods or technique which can help in enhancing the climate change resilience of the local communities. Environmental isotopes such as Oxygen-18, Deuterium (^{2}H), and tritium (^{3}H) and isotopes are useful to identify the origin of recharge, recharge area. It is also helpful to identify the source of contamination of spring water.

Acknowledgements Authors are grateful to the Department of Agricultural Engineering, North Eastern Regional Institute of Science and Technology, Nirjuli, Arunachal Pradesh, for providing the support and valuable suggestions for bringing this paper.

References

Agarwal A (2000) Drought? Try capturing the rain. Center Sci Environ 1–16
Agarwal A, Agarwal NK, Rathi VK (2015) Water stress on springs of Lesser Himalayan region. British J Appl Sci Technol 9(3):243–255
Agarwal A, Bhatnaga NK, Nema RK, Agrawal NK (2012) Rainfall dependence of springs in the Midwestern Himalayan Hills of Uttarakhand. Mt Res Dev 32(4):446–455
Barbieri M (2019) Isotopes in hydrology and hydrogeology. Water 11:291
Bhat NA, Jeelani G (2015) Delineation of the recharge areas and distinguishing the sources of karst springs in Bringi watershed, Kashmir Himalayas using hydrochemistry and environmental isotopes. J Earth Syst Sci 124(8):1667–1676
Clark ID, Fritz P (2013) Environmental isotopes in hydrogeology. CRC Press
Craig H (1961) Isotopic variations in meteoric waters. Science 133(3465):1702–1703
Dhakal D, Tiwari A, Tambe S, Sinha UK, Arrawatia ML (2014) Isotope studies to identify the origin and recharge area of Himalayan springs as a climate change adaptation initiative: a case study from Sikkim, Eastern Himalaya. Int J Earth Sci Eng 07(01):136–140
Fontes JC, Letolle R, Olive P, Blavoux B (1967) Oxygene 18 et tritium dans le basin d'Evian. In: Proceedings of symposium on isotopes in hydrology. International Atomic Energy Agency, Vienna, p 401
Giberson D, Knysh K, MacDonald E (2013) Springs of the clearest and purest water. Island Mag 73:2–11
Jeelani G (2008) Aquifer response to regional climate variability in a part of Kashmir Himalaya in India. Hydrogeol J 16(8):1625–1633
Jeelani GH, Bhat NA, Shivanna K (2010) Use of δ18O tracer to identify stream and spring origins of a mountainous catchment: A case study from Liddar watershed, Western Himalaya. India. J Hydrol 393(3–4):257–264

Jeelani G, Shah RA (2014) Distinguishing and estimating the sources of spring recharge: a case study of Martand Karst Spring of Kashmir Valley
Jitendra (2015). Down to earth. https://www.downtoearth.org.in/news/agriculture/about-100-districts-have-suffered-drought-conditions-for-15-years-50993.
Kumar CP (2013) Hydrological studies using isotopes. Int J Innov Res Dev 2(13)
Kumar R, Singh RD, Sharma KD (2005) Water resources of India. Curr Sci 794–811
Kusre BC, Bora PK, Rai D, Adhikari K, Niranjit Khuman C, Tabing J (2018) Developing water resources management strategies for South Sikkim District, India. Water Sci Technol Water Supply 18(2):648–659
Kusre BC, Bora PK, Rai D, Suranjoy Singh S, Tana A, Subba S (2017) Combating water scarcity through roof water harvesting: planning and design with stakeholders' perception in Sikkim (India). Water Sci Technol Water Supply 17(3):799–810
Lee KS, Wenner DB, Lee I (1999) Using H-and O-isotopic data for estimating the relative contributions of rainy and dry season precipitation to groundwater: example from Cheju Island. Korea. J Hydrol 222(1–4):65–74
Lepcha P T, Pandey PK, Ranjan P (2021) Hydrological significance of Himalayan surface water and its management considering anthropogenic and climate change aspects. In: IOP conference series: materials science and engineering, vol 1020, No. 1, p 012013
Mahamuni K, Upasani D (2011) Springs: a common source of a common resource. In: 13th Biennial conference of the international association for the study of the commons (IASC), Sustaining commons: sustaining our future.
Negi GCS, Joshi V (2002) Drinking water issues and development of spring sanctuaries in a mountain watershed in the Indian Himalaya. Mt Res Dev 22(1):29–31
NITI Aayog (2017) Inventory and revival of springs in the himalayas for water security. Depertment of Science and Technology, Government of India, New Delhi
Price RM, Swart PK (2006) Geochemical indicators of groundwater recharge in the surficial aquifer system, Everglades National Park, Florida, USA. Geol Soc Am Spec Pap 404:251–266
Ranjan P, Pandey PK (2020) Reviving, development and protection of springs to increase water security in the Himalayan Region. In: Proceedings of the 5th international conference on computers & management skills, pp 25–29
Rees GH, Collins DN (2006) Regional differences in response of flow in glacier-fed Himalayan rivers to climate warming. Hydrol Process 20:2157–2167
Rozanski K, Araguás-Araguás L, Gonfiantini R (1993) Isotopic patterns in modern global precipitation. Clim Change Continental Isotopic Rec 78:1–36
Shah RA, Jeelani G, Jacob N (2017) Estimating mean residence time of karst groundwater in mountainous catchments of Western Himalaya, India. Hydrol Sci J 62(8):1230–1242
Sidle WC (1998) Environmental isotopes for resolution of hydrology problems. Environ Monit Assess 52(3):389–410
Singh P, Kumar V, Thomas T, Arora M (2008) Changes in rainfall and relative humidity in river basins in northwest and central India. Hydrol Process 22:2982–2992
Stahl W, Aust H, Dounas A (1974) Origin of artesian and thermal waters determined by oxygen, hydrogen and carbon isotope analyses of water samples from the Sperkhios Valley, Greece. In: Proceedings of a symposium on isotope techniques in groundwater hydrology, vol I
Tambe S, Kharel G, Arrawatia ML, Kulkarni H, Mahamuni K, Ganeriwala AK (2012) Reviving dying springs: climate change adaptation experiments from the Sikkim Himalaya. Mt Res Dev 32(1):62–72
Tambe S, Kharel G, Subba S, Arrawatia ML (2013) Rural water security in the sikkim Himalaya: status, initiatives and future strategy. India Mountain Initiative, Summit in Kohima, Nagaland
Vashisht AK (2015) Adaption of seepage spring development technique to manage the water scarcity in Himalayan region–a practical approach. Water Util J 11:93–98
Vashisht AK (2017) Use of mathematical techniques for determining the reliability of springs. In: Development of water resources in India. Springer, Cham, pp 113–124

Vashisht AK, Ranjan P (2020) Intermittent Multi-column sand filter: a unique solution to multiple applications. J Inst Eng (India) Ser A 101(1):69–75
Vashisht AK, Sharma HC (2007) Study on hydrological behaviour of a natural spring. Curr Sci, 837–840

Chapter 10
Water Hammer Analysis for Pipe Line Network Using HAMMER V8i

Ajmal Hussain, Muhammad Mustafa, S. M. Ahbar Warsi, and Sumit Kumar

Abstract Hydraulic transients occur as a direct result of rapid variations of flow field in pressurized (closed-conduit) systems. The change in velocity from valve closures or pump operations causes pressure surges that are propagated away from thesource throughout the pipeline. If the maximum pressures exceed the bar ratings(mechanical strength) of the piping material, different types of failure such as pipe bursts canoccur. Similarly, if the minimum pressure drops below the vapour pressure of the fluid, cavitationcan occur and can be detrimental to the pipeline system. The purpose of present study is to asses and simulate the hydraulic transients in a pipe line network of treated effluent rising main of Mpophomeni sanitation scheme using Bentley HAMMER V8i. A total of five scenarios were simulated using different combinations. The simulation results shows that the transient pressures in the pipeline exceeded the bar rating of the pipe where the burstsor cavitation may occur for the simulated scenarios. This study shows that the transient pressures in pipe line system were reduced to safe limit after providing water hammer protection devices.

Keywords Hydraulic transient · Water hammer · Cavitations · Bentley HAMMER V8i

10.1 Introduction

Water hammer commonly occurs when a valve closes suddenly at an end of a pipeline system, and a pressure wave propagates in the pipe (Chaudhry 1979). Under steady state conditions in a pipeline system, flow variables like discharge remainconstant. However, if a sudden change occurs in the system through a change in control operations such asthe closure of an outlet valve or the sudden shutdown of a pump due

A. Hussain (✉) · S. M. A. Warsi · S. Kumar
Department of Civil Engineering, Zakir Hussain College of Engineering and Technology, AMU, Aligarh 202002, India

M. Mustafa
Design Section, SMEC India Private Limited, Gurgaon, India

R. Jha et al. (eds.), *Water Resources Management and Reservoir Operation*, Water Science and Technology Library 107,
https://doi.org/10.1007/978-3-030-79400-2_10

to power failure, a transientstate is initiated, and it takes a finite amount of time before another (new) steady-state conditionis established in the pipeline system. The flow phenomenon associated with such rapid changesis called a hydraulic (or fluid) transient. The main concern during a hydraulic transient in asystem is the rapid fluctuations in the pressure since dramatic changes in the pressure can resultin catastrophic damage to pipelines and hydraulic machinery.

By closing the valve rapidly,the valve converts the kinetic energy carried by the fluid particles into strain energy in the pipewalls. This results in a "pulse wave" of abnormal pressure to travel from the disturbance into thepipe system. Energy losses due to mainly friction cause the transient pressure waves todecay until a new steady state is established. Hydraulic transient events in water distribution system can cause significant damage, disruption,and expense (Huo et al. 2007). In general, transient events are usually most severe at control valves, pump stations, in high-elevation areas, and in remote locations that are far from overhead storage tanks. However, all systems have to start up, switch off, undergo flow changes, and so on. In addition, water systems are not immune from human errors, malfunction and break downof mechanical devices, and other risky events. Emadi and Solemani (2011) investigated the effect of parameters such as pipe diameter, thickness and type, moment of inertia and temperature on maximum water hammer in the context of Kuhrang Pumping Station. Fluid pipeline failures due to water hammer effects are described in detail by Schmitt et al. (2006).

During a hydraulic transient state, a pipeline may be subjected to objectionable high and lowpressure cycles. The high pressures can damage the pipeline system components, such as valves,pumps, and other pipeline components, as discussed earlier. The change in the fluid velocity (more correctly discharge) in the pipeline systems is the first step that leads to a hydraulictransient. The resulting change in pressure is directly proportional to the change in velocity. Hence, as much as possible, sudden changes in the velocity should be avoided to minimize theoccurrence of pressure transients in the system. Bergant et al. (2012) presented a comprehensive water hammer analysis of pumping system for control of water in underground mines. Deshmukh (2014) presented the hydraulic transient analysis of Kolar water pipeline using using Bentley HAMMER V8i.

The present study was conducted using the popular surge software Bentley HAMMER V8*i*. Bentley HAMMER V8*i* is a versatile program capable of modeling any type of surge protection device and its powerful graphical results presentation and interpretation capability has helped thousands of engineers worldwide design large complex transmission mains, small branching networks as well as large distribution networks for over 30 years. Bentley HAMMER V8*i* is based on technology originally created by Environmental Hydraulics Group (HAMMERTM 2005). Water hammer equations for elastic pipes produce a 1-D partial equation and may be solved by Methods of Characteristics (MOC). Hammer Software uses the Method of Characteristicto solve non-linear differential equations which have the following form, Evangelisti (1969), Fox (1977), Streeter (1967, 1972):

$$\frac{dV}{dt} + \frac{1}{\rho c} \cdot \frac{dP}{dt} + \frac{f|V|V}{2D} = 0 \tag{10.1}$$

$$a^2 \frac{\partial V}{\partial s} + \frac{1}{\rho} \frac{\partial \rho}{\partial t} = 0 \tag{10.2}$$

Solution of the above equations using MOC will be

C+:

$$\frac{a}{gA}(Q_p - Q_{i-1}) + (H_p - H_{i-1}) + \frac{f \Delta x}{2gDA^2} Q_p |Q_{i-1}| = 0 \tag{10.3}$$

C-:

$$\frac{a}{gA}(Q_p - Q_{i+1}) - (H_p - H_{i+1}) + \frac{f \Delta x}{2gDA^2} Q_p |Q_{i+1}| = 0 \tag{10.4}$$

For the solution there are many boundary conditions are considered requiring like reservoirs, pumps, pipeline branches, dead ends etc. This method not only saves times but also reduces likelihood of mistakes which may occur while copying date to the software.

10.2 Problem Statement and Procedure of Analysis

The present study is carried for a pipe line network of treated effluent rising main of Mpophmeni sanitation scheme, South Africa which consists of following major units:

a. Pump: Pump working with the capacity of supplying a flow rate of 250 m^3/h up to a head of 76 m. Pump has a rated power of 90 kW at 1450 rpm.
b. Rising Main Network: Rising main consists of PVC class 12 pipe of diameter 400 mm for first 3000 m and then the pipe diameter reduces to 315 mm upto 5423 m,then pipe diameter increases to 400 mm till 5503 m,then again reduces to 315 mm and continues till the tail end discharging into the receiving chamber. The whole network configuaration is shown in Fig. 10.1.

Profile for the elevation with chainage of the rising main network is shown in Fig. 10.2. The pump station is located at zero chainage of the profile and the rising main culminates in free fall at a receiving chamber at the tail end. As evident from Fig. 10.2; three major peaks and low points are observed along the network path and in the last leg of the network i.e. after chainage of around 5500 m, network will be acting under gravity. The undulating topography of the rising main path exposes the network to the risk of surge occurrences in events of power failure, sudden closure of valves etc. The main objective of the study is to identify transient issues for this system and recommend surge protection alternatives.

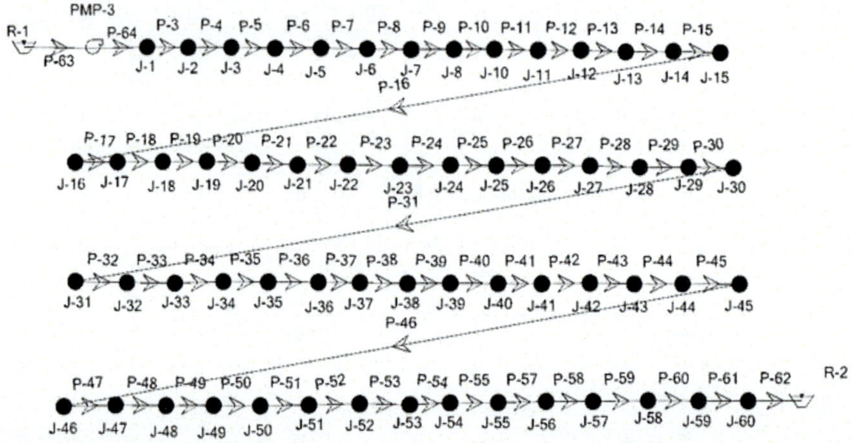

Fig. 10.1 Steady state model for problem 2

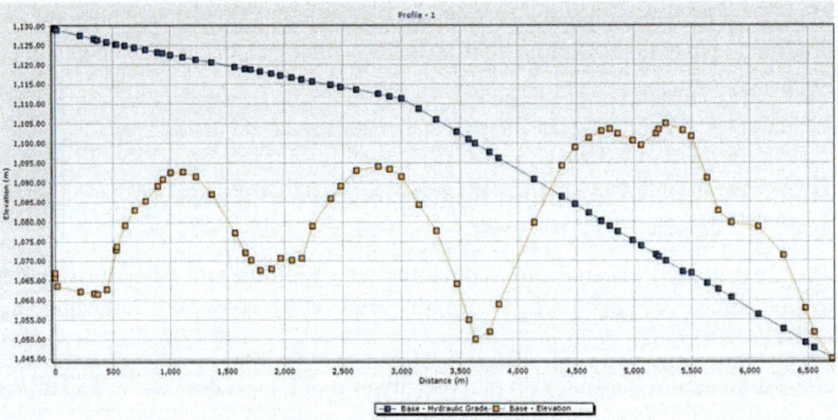

Fig. 10.2 Profile of hydraulic grade line and elevation with chainage of the rising main network

10.3 Results and Analysis

10.3.1 Baseline Scenario: Steady State Conditions

A baseline run of the network is conducted to identify the baseline scenario that is network running under steady state without any transient event. This is conducted to identify the steady state conditions for the network. Baseline scenario model is shown in Fig. 10.2.

Figure 10.2 indicates thatHGL is significantly below the ground elevation after around 4300 m chainage. So, to counter negative pressure heads in this region, a

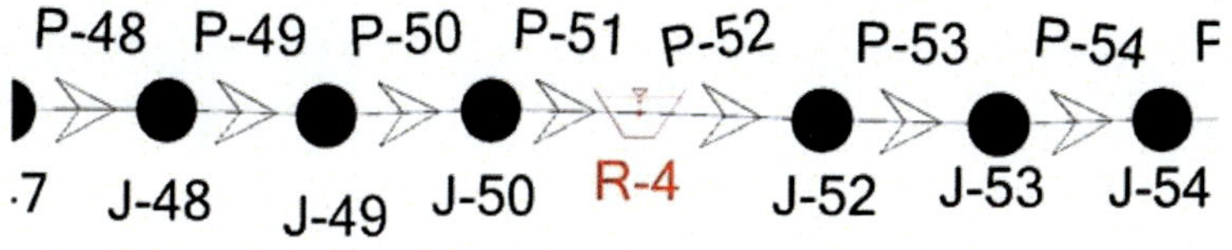

Fig. 10.3 Reservoir location between J-50 and J-52

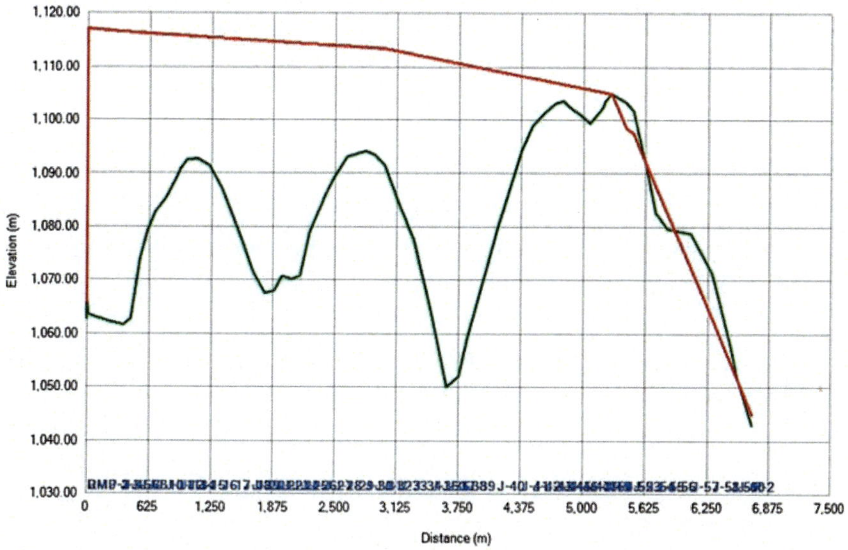

Fig. 10.4 Hydraulic Grade Line after providing the reservoir for the network

reservoir at the highest elevation point(1105 m) at J51 is provided which is shown in Fig. 10.3.

After providing the reservoir; Hydraulic Grade Line (HGL) for the network is shown in Fig. 10.4. As a result the negative pressure heads are well within the limits ie −10 m H_2O (or −98.1 kPa) and the positive pressure heads are also within the safe limits as shown in Fig. 10.5.

10.3.2 *Surge Analysis on Baseline Network Without Surge Protection*

In the next stage the impact of a power failure is simulated without any surge protection device. Hydraulic Grade Line for baseline network without surge protection device is shown in Fig. 10.6. For the analysis it is assumed that the check valves installed at the pump closes after 5 s of power failure, which is below the critical

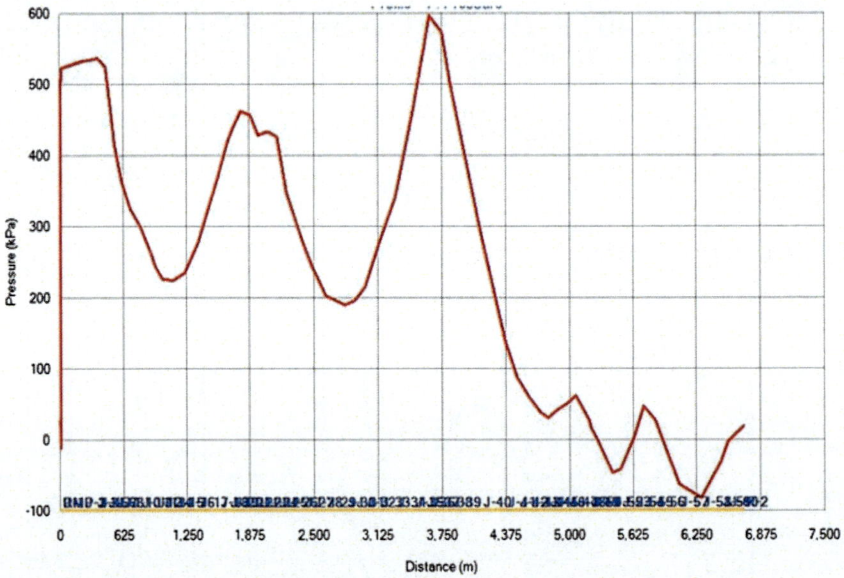

Fig. 10.5 Pressure diagram for steady state analysis after providing reservoir

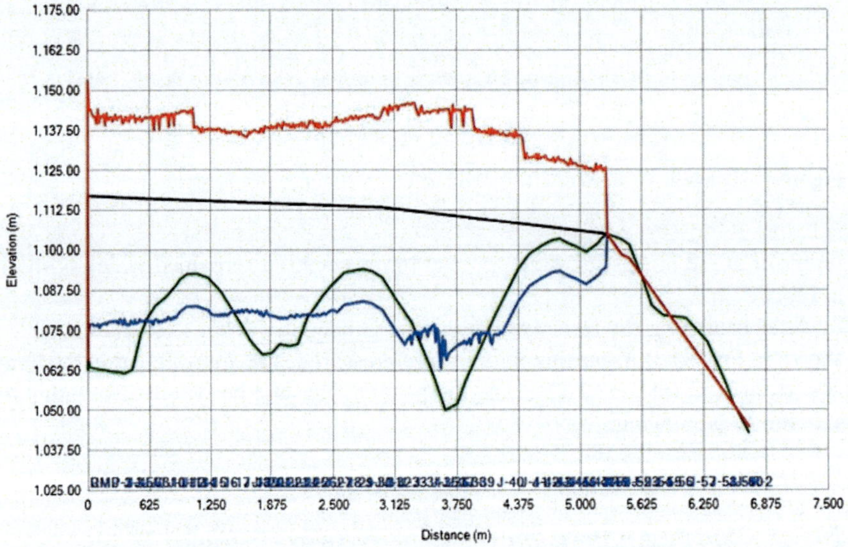

Fig. 10.6 Hydraulic grade line for baseline network without surge protection device

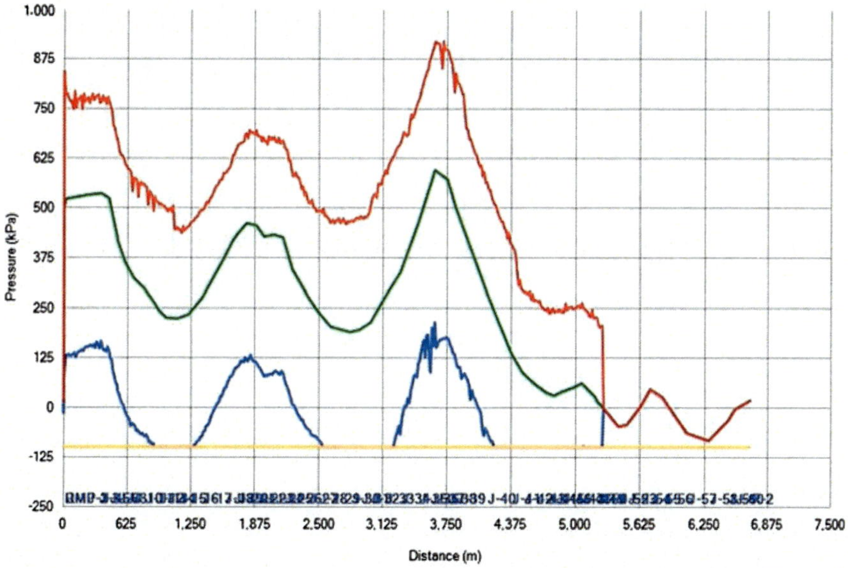

Fig. 10.7 Pressure diagram for transient analysis for the network without surge protection device

time period. After running transient simulation, it's found that the transient effect is only in the portion before the reservoir at high point as shown in Figs. 10.6 and 10.7. Also negative pressures are below −98 kPa in many parts of the pipeline. So, surge protection devices are required for protection from water hammer in this region only.

10.3.3 Analysis with Surge Protection Device

10.3.3.1 With Application of Four Air Valves

To minimize negative pressure heads, double acting Air Valves, with inflow orifice dia. 80 mm and out flow orifice dia. 2.0 mm, are adopted, at different locations shown in Table 10.1 and Fig. 10.8.

The results obtained after adding valves in the pipe line network are shown in Fig. 10.9. The locations at which valves were added, negative pressures are reduced

Table 10.1 Location of air valve at different chainage along the pipe line

Air valve no	Chainage
AV-1	1080
AV-2	2728
AV-3	4785
AV-4	5260

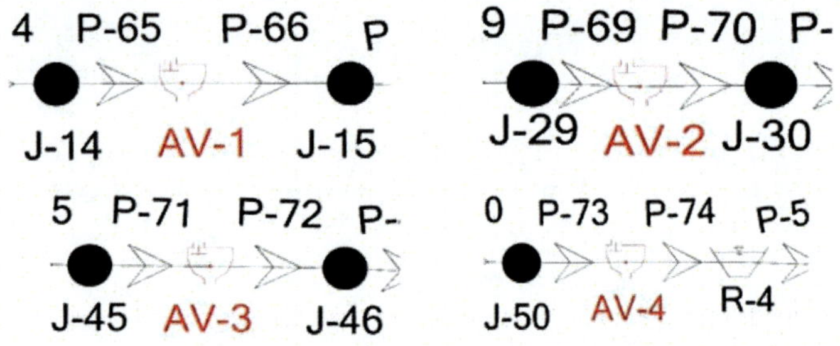

Fig. 10.8 Air valve location (AV-1, AV-2, AV-3 & AV-4) at different locations of the network

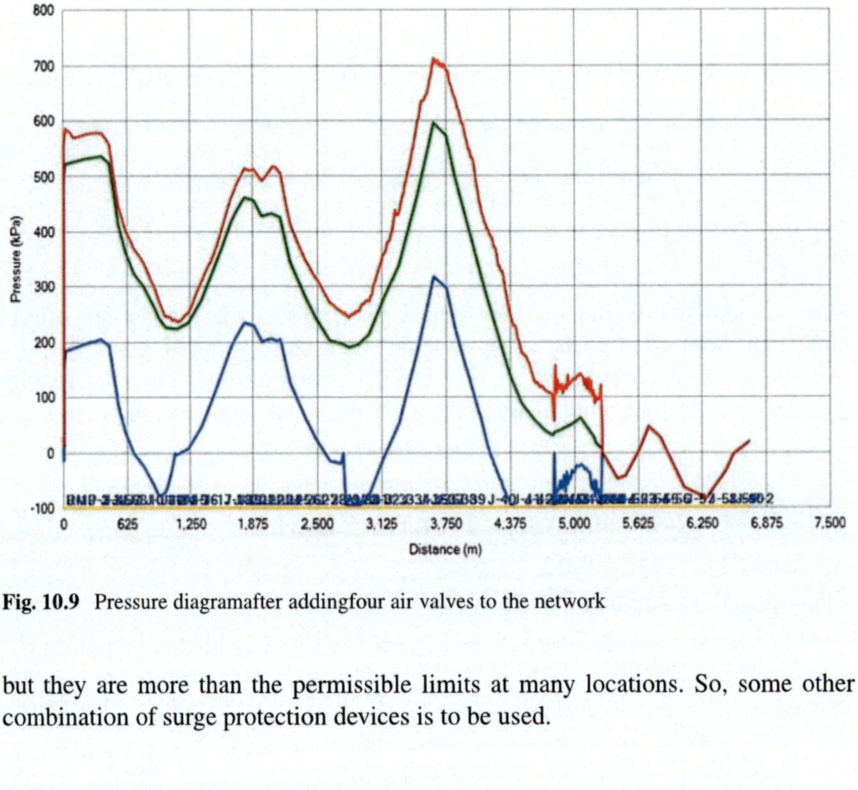

Fig. 10.9 Pressure diagramafter addingfour air valves to the network

but they are more than the permissible limits at many locations. So, some other combination of surge protection devices is to be used.

10.3.3.2 With Application of 4 Air Valves and 1 Hydropneumatic Tank

To further reduce the negative pressures, a hydropneumatic tank to the network at J1 is provided as shown in Fig. 10.10. The properties of the hydropneumatic tank are:Volume = 2000 L,Liquid Volume(Initial) = 1600 L, Tank Calculation Model = Gas Law Model, Dia. (Tank inlet Orifice) = 175 mm and HGL (initial) = 1140 m.

The results obtained after adding four valves and a hydropneumatic tank in the pipe line system is shown in Figs. 10.11 and 10.12. After adding hydropneumatic

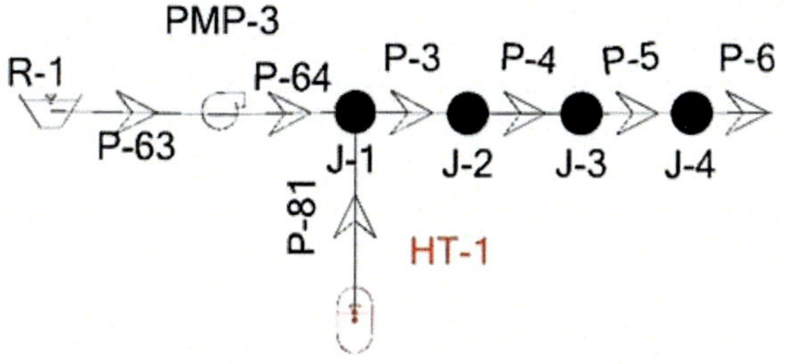

Fig. 10.10 Hydropneumatic tank at J1

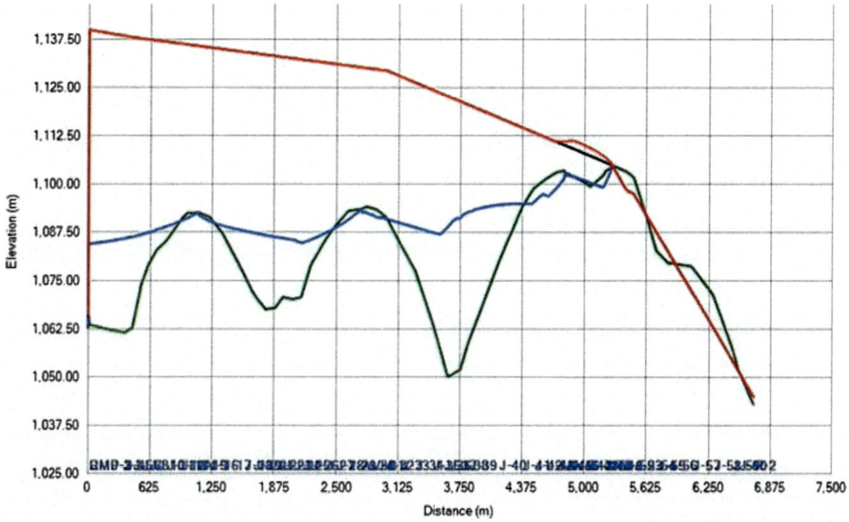

Fig. 10.11 Hydraulic grade line with elevation and chainage for the network with hydropneumatic tank and four air valves

tank, the negative pressure values have been reduced significantly and are within safe limits as shown in Fig. 10.12. Positive pressures also do not exceed the safe limits. Therefore, it can be said that the pipe line network of treated effluent rising main of Mpophmeni sanitation scheme is safe with the present mitigation measures.

10.4 Conclusion

Based on the findings, application of four air valves and one hydropneumatic tank is recommended for the safe operation of treated effluent rising main of Mpophmeni sanitation scheme. Without surge protection device, the negative pressure was greater

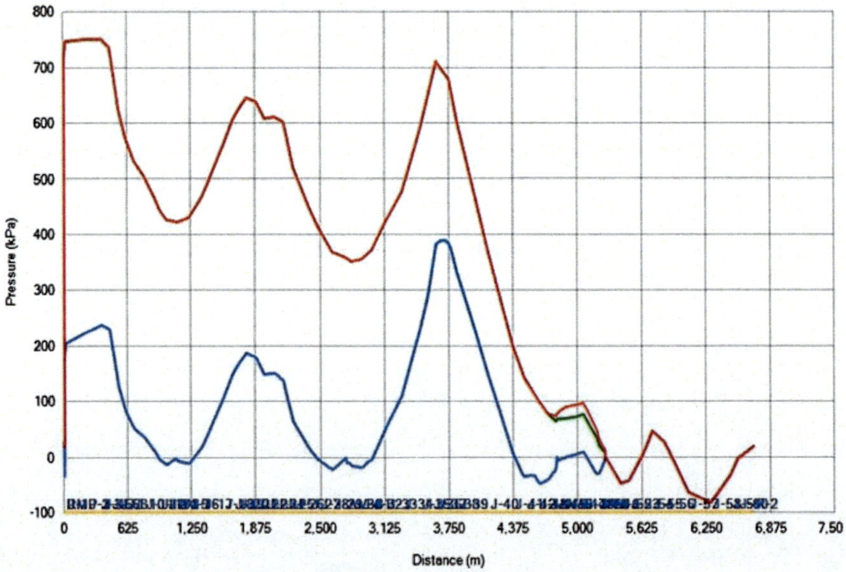

Fig. 10.12 Pressure diagram for transient analysis with hydropneumatic tank and four air valves

than −98 kPa on several points, but after providing surge protection, negative pressures are well below the safe limit. The hydropneumatic tank may be provided at the immediate downstream of the pump. A minimum of above four air valves shall be provided to contain the effect of downsurge in the network, however any additional air valves provided in the network will further improve the network performance. It can be concluded that the present developed model for the pipeline system reduces the risk of damages associated with water hammer and consequently increase the safety and as well as reduce the failure rate for the present pipe line system. It also reduces wearing and tearing effects of water hammer in pumping and pipeline systems, and increase lifetime of the infrastructure.

References

Bergant A, Simpson AR, Sijamhodzic E (2012) Water hammer analysis of pumping system for control of water in underground mines. International mine water association, pp 9–19

Chaudhry MH (1979) Applied hydraulic transients. Van Nostrand Reinhold Company, New York

Deshmukh TS (2014) Hydraulic transient analysis of kolar water pipeline using Bentley HAMMER V8i. Int J IJERT 3(9)

Emadi J, Solemani A (2011) Maximum water hammer sensitivity analysis. IJERT

Evangelisti G (1969) Waterhammer analysis by the method of characteristics. L'Energia Elettrica, Nos. 10–12

Fox JA (1977) Hydraulic analysis of unsteady flow in pipe networks. The Macmillan Press Ltd., London and Basingstoke

HAMMERTM User Guide, Bentley Systems Inc. (2005)
Huo J, Eckmann DH, Morris KE (2007) Surge protections: review, analysis, and engineering design. In: Proceedings of AWWA/ACE07, Toronto, Ont., Canada
Schmitt C, Pluvinage G, Hadj-Taïeb E, Akid R (2006) Water pipeline failure due to water hammer effects. Fatigue Fract Eng Mater Struct 29(12):1075–1082
Streeter VL (1967) Water-hammer analysis of distribution systems. J Hydraulics Div Proc ASCE 93(HY5):185–201
Streeter VL (1972) Unsteady flow calculations by numerical methods. Proc ASME J Basic Eng 94(series D, No. 2):457–466

Chapter 11
Dam Break Flood Routing and Inundation Mapping Using HEC-RAS and HEC-GeoRAS

A. Bharath, Anand V. Shivapur, and C. G. Hiremath

Abstract Dam break analysis helps to predict the breach flood hydrograph and its time of arrival at desired locations along with depth and velocity of flow. This paper illustrates a case study on dam break analysis of Hidkal Dam using one-dimensional hydraulic model HEC-RAS with the help of river geometry generated by DEM. The present study illustrates the simulation of flood hydrograph under PMF scenario for both piping and overflow failure modes. The peak flow values as obtained from the model are 72,020.57 m^3/s and 78,384.21 m^3/s at the downstream of the dam and the corresponding inundation area estimated to be of 74.32 km^2 and 78.19 km^2 for piping failure and overtopping failure, respectively. The flow depth is found to vary from 30.39 m to 13.09 m for piping failure and 34.98 m to 13.86 m for overtopping failure at the downstream of the dam and the end of river reach, respectively.

Keywords Dam break analysis · Breach flood hydrograph · Breach parameters · HEC-RAS · Flood inundation

11.1 Introduction

A dam is a barrier constructed across a river, which obstructs or directs the flow and often creates a reservoir or lake. The water impounded by the dam is used for many purposes like drinking water supply, irrigation water supply and hydroelectric power generation. Though the dams have many benefits for our society, their failures cause most devastating disasters. Dam failures are very rare events, but it causes immense damage to life and property, hence it generally reminds engineers the probability of dam failure (Abhijith et al. 2017). Dam break can be summarized as partial or catastrophic failure of a dam resulting in uncontrolled flood wave of greater

A. Bharath (✉)
Department of Civil Engineering, GITAM University, Bangalore, India

A. V. Shivapur · C. G. Hiremath
Department of Water and Land Management, Centre for PG Studies, VTU, Belagavi, Karnataka, India

R. Jha et al. (eds.), *Water Resources Management and Reservoir Operation*,
Water Science and Technology Library 107,
https://doi.org/10.1007/978-3-030-79400-2_11

height traveling at a very high speed which can cause damage to life, property and infrastructure in the downstream (Xiong 2011).

Piping, extreme storm, landslide, earthquake, malfunctioning of equipment, foundation failure, sabotage, structural damage are the mechanisms that can cause dam failures (USACE 2016). Regardless of the reason, majority of the dam failures begin with the breach formation (Mohamed 2018; Wang et al. 2015).

A breach can be defined as the opening formed in the dam which results in dam failure and this event produces a flood wave from the water impounded behind the dam. Though the main failure modes have been recognized as overtopping or piping, the real failure mechanics for both earthen and concrete dams are not well understood (Xiong 2011; Zhang et al. 2009).

Dam break analysis helps in predicting the peak flood and areas affected by flood in the downstream due to dam breach which helps in the estimation of rehabilitation cost. Such analysis predicts the potential of precautionary measures which can be taken to avoid the dam break which minimizes or avoid damage (Kulkarni et al. 2016). Dam break modeling helps to simulate dam break flood wave propogation along the downstream of the dam (Khosravi et al. 2019).

In the present study, a dam break analysis has been carried out using a hydraulic model HEC-RAS for different failure mode to predict the breach flood, floodplain mapping as suggested by Abhijith et al. (2017), Ackerman et al. (2008), Xiong (2011).

11.2 Study Area

Hidkal dam is selected for dam break analysis as no such studies have been reported on it. It is constructed across Ghataprabha River, and is located in Hidkal village, Belagavi District, Karnataka, India. The Hidkal dam has a main section of length 4481 m and total length is 10,183 m. The spillway section is of masonry structure whereas the remaining section is of earthen embankment. The height of the dam is 53.34 m above the foundation and it has a storage capacity of 51 TMC ft irrigating an area of 317,435.41 ha.

11.3 Methodology

In the present study, hydraulic model HEC-RAS has been used for dam break analysis and extraction of river geometry data and floodplain mapping has been done by HEC-GeoRAS. HEC-RAS uses St. Venant's equations for flood routing for unsteady flow (Ackerman and Brunner 2008; Derdous et al. 2015). Figure 11.1 represents the workflow processes for dam break analysis and Flood Plain mapping using HEC-RAS and HEC-GeoRAS.

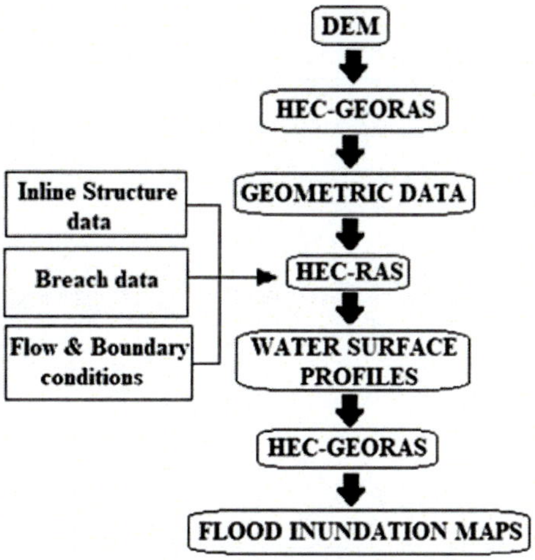

Fig. 11.1 Workflow processes for dam break analysis and flood plain mapping

Data requirement and processing

The following data are used for the analysis:

- Salient features of the dam in study reach of the river.
- Probable Maximum Flood.
- Elevation-storage/area relationship of the reservoir.
- Cross sections of the river from dam site to the most downstream location of interest.
- Manning's roughness coefficient for the river.
- Breach Data.
- Boundary conditions.

Salient features of the dam, probable maximum flood and elevation-storage/area relationship of the reservoir are collected from dam authority. River geometry is extracted with the help of HEC-GeoRAS tool using DEM data (Source Earth Explorer Website (earthexplorer.usgs.gov) 32 m resolution). Through visual inspection of the site using web imagery, the manning's roughness coefficients are chosen based on land use characteristics (Bharath 2020a, b). The breach parameters are final bottom breach width, depth, side slope and formation time. Froehlich (2008) regression equations are used to determine the breach parameters (Bharath 2020a, b). Boundary conditions greatly influences the downstream floodwater, hence they must be carefully selected and defined. In this analysis, the boundary conditions defined are; flow hydrograph (i.e., PMF) at upstream, normal depth at downstream and time-series of gate opening at inline structure.

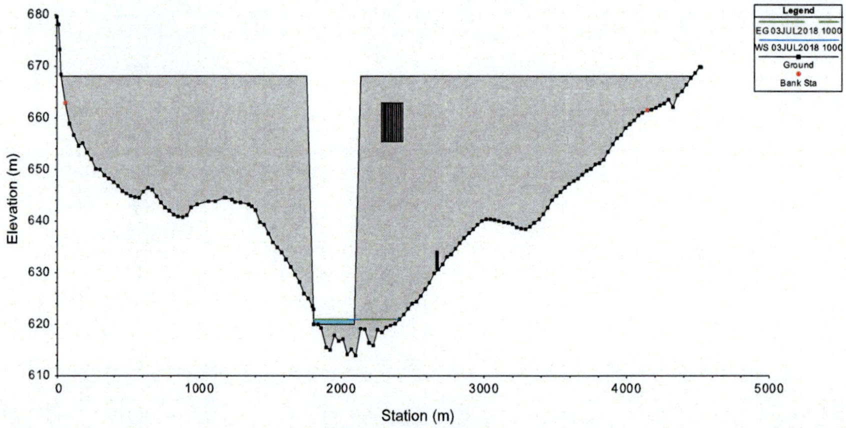

Fig. 11.2 Dam section at the end of the breach simulation

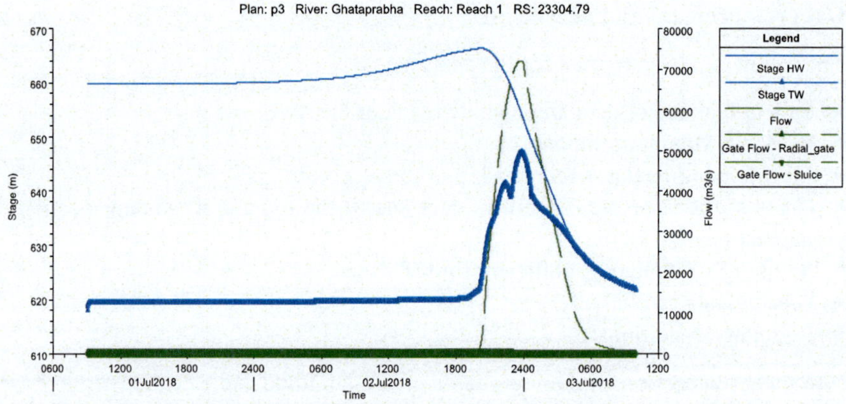

Fig. 11.3 Stage-flow hydrograph at the inline structure for piping failure

11.4 Results and Discussions

The dam break analysis is performed for unsteady flow condition for piping and overtopping failure considering reservoir full condition. The river geometry is extracted for 22 km stretch in the downstream; the cross section is extracted and interpolated to a closer interval of 150–200 m. The simulation is done considering that the PMF starts at 1 July 2018 at 9 am and ends at 3 July 2018 at 10 am and hence results obtained are corresponding to the simulation period.

Figure 11.2 shows the dam profile at the end of the simulation, it shows the breach formed on the dam and depleted water level at the end of the simulation.

Figures 11.3 and 11.4 show the headwater and tail water stage hydrograph, flood hydrograph and discharge through the gates at the inline structure for piping and

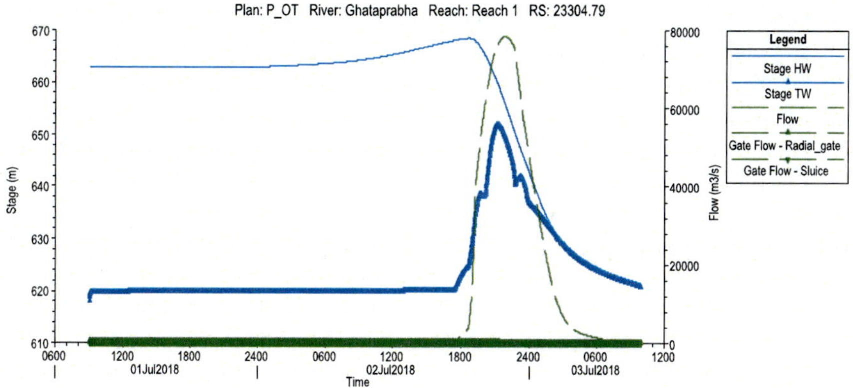

Fig. 11.4 Stage-flow hydrograph at the inline structure for overtopping failure

overtopping failure modes, respectively. Table 11.1 shows the maximum values of headwater stage, tailwater stage and flow rate and their arrival time.

Flood inundation maps

Floodplain maps are produced in HEC-GeoRAS platform using HEC-RAS results. Figure 11.5 shows the area inundated due to piping and overtopping mode of failure.

Table 11.1 Maximum stage-flow details

Particular	Piping failure		Overtopping	
	Maximum	Time at maximum	Maximum	Time at maximum
HW Stage (m)	665.91	02-07-18 20:14	667.85	02-07-18 18:40
TW Stage (m)	646.51	02-07-18 23:48	651.17	02-07-18 21:14
Flow (m^3/s)	72,020.57	02-07-18 23:36	78,384.21	02-07-18 21:50

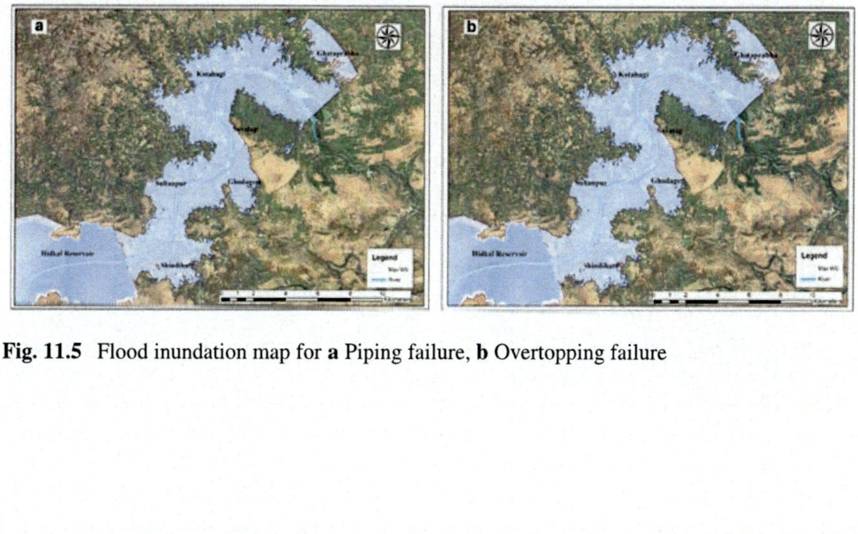

Fig. 11.5 Flood inundation map for **a** Piping failure, **b** Overtopping failure

It is found that the area of inundation is 74.32 km^2 and 78.19 km^2 due to piping and overtopping failure, respectively.

Figure 11.5 depicts the major villages that gets affected due to dam failure. Totally, twenty villages get affected among which seventeen are severely affected. These villages have a great threat of loss of life and property. The majority of the area that is being affected is agricultural land hence there may be a huge loss of crops which in turn may affect the economy.

Flow variation at different river stations

Flood hydrographs at different downstream stations (downstream of the dam, 5, 10, 15, 20 and 22 km away from the dam) are compared as shown in Figs. 11.6 and 11.7. The results obtained depicts that the peak flow value is very high at the station closer to the dam and it gradually decreases as the distance increases. The kinks formed

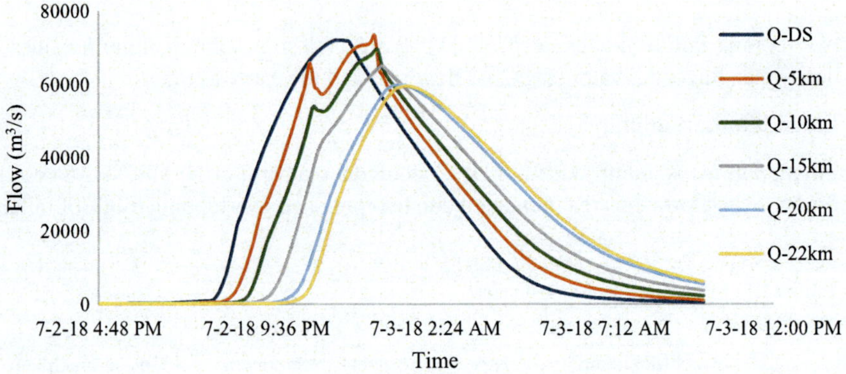

Fig. 11.6 Flood hydrographs for piping failure

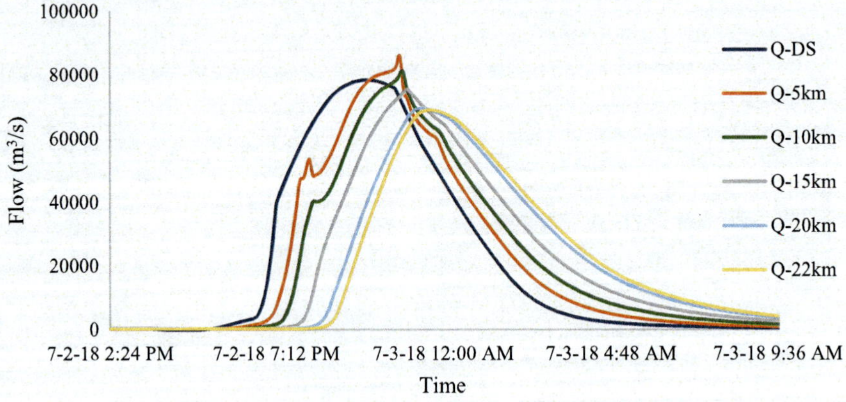

Fig. 11.7 Flood hydrographs for overtopping failure

in the peak flood hydrographs are may be due to surge created in the flow due to meandering of streams and valley cross section.

Stage variation at different river stations

Stage hydrographs at different stations (upstream & downstream of the dam, 5, 10, 15, 20 and 22 km away from the dam) are compared as shown in Figs. 11.8 and 11.9. The results obtained depicts that the peak stage value is very high at the station closer to the dam and it gradually decreases as the distance increases. The flow depth is found to vary from 30.39 m to 34.98 m for piping failure and 34.98 m to 13.86 m

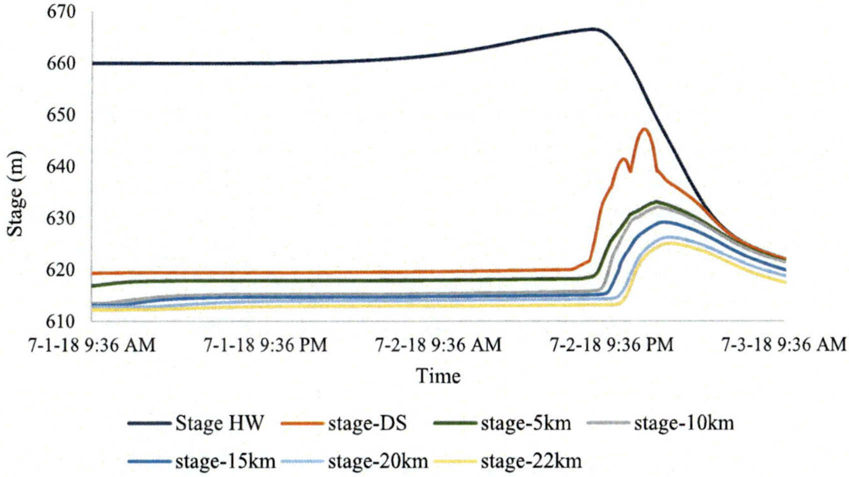

Fig. 11.8 Stage hydrographs for piping failure

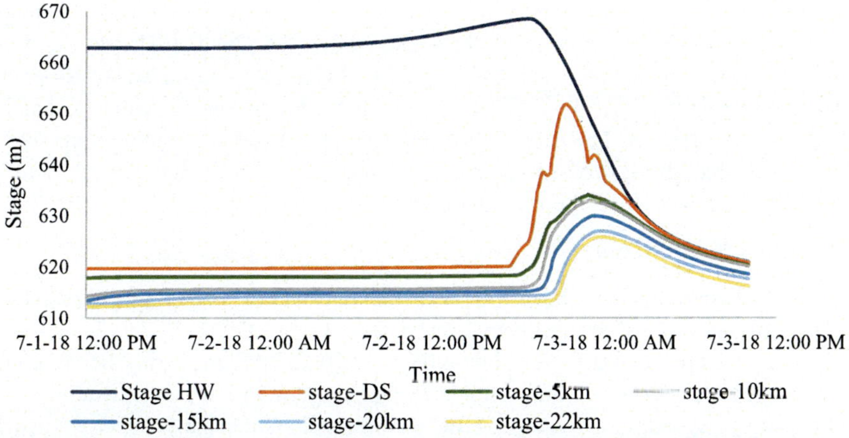

Fig. 11.9 Stage hydrographs for overtopping failure

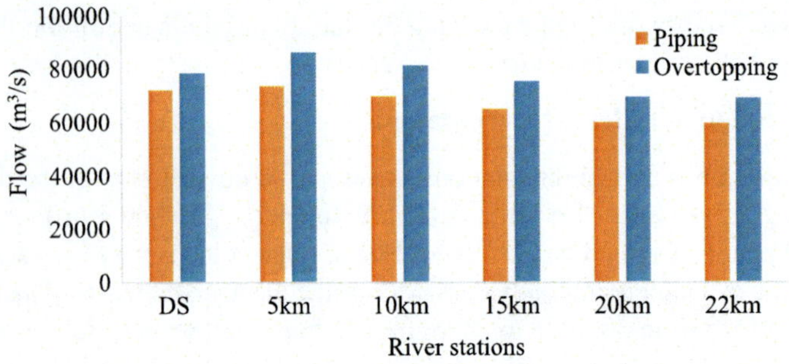

Fig. 11.10 Peak flow magnitudes at different river stations

for overtopping failure at the downstream of the dam and the end of river reach, respectively. The kinks formed in the stage hydrographs at the downstream station are may be due to surge formed behind the dam during failure.

Figure 11.10 highlights the peak flow values of piping and overtopping failures at different river stations. This comparison depicts that peak flow due to overtopping failure is higher than that of piping failure at all the river station hence overtopping is considered as the most critical failure mode for dam break analysis.

11.5 Conclusions

Dam break model is simulated for the worse scenario, i.e., unsteady flow condition (PMF) and reservoir full condition for both piping and overtopping failure considering reservoir full condition. HEC-RAS gives an easy platform to route the dam breach flood along the downstream of the river and helps in determining the hydraulic conditions at critical locations and different river stations. Floodplain maps are generated in HEC-GeoRAS platform using HEC-RAS results. The results show that the area of inundation is 74.32 km^2 and 78.19 km^2 due to piping and overtopping failure, respectively. Totally, twenty villages get affected among which seventeen are severely affected. The obtained results can be utilized in planning dam safety measures, flood mitigation measures, land use planning in the downstream and to develop an emergency response plan during the flood.

Limitations of the study

DEM data used for generating the cross section data was of coarser resolution which affects the accuracy of results given from the model hence DEM of finer resolution or cross sections from field survey should be used. HEC-RAS one-dimensional model was used which gives one-dimensional results hence it will not work where multi-dimensional modeling is required. HEC-RAS model assumes flow is perpendicular to the cross sections which might result in improper inundation area when the flow increases.

Acknowledgements The authors would like to take this opportunity to recognize the efforts and assistance from every individual on all aspects in the preparation of this paper.

References

Abhijith R, Amrutha G, Vijayaraj G, Rijisha TV (2017) Dam break analysis of Idukki Dam using HEC RAS. Int Res J Eng Technol (IRJET) 4(7):3410–3415

Ackerman CT, Brunner GW (2008) Dam failure analysis using HEC-RAS and HEC-GeoRAS. In World environmental and water resources congress 2008: Ahupua'A, 8

Ackerman CT, Fleming MJ, Brunner GW (2008) Hydrologic and hydraulic models for performing dam break studies. In :World environmental and water resources congress 2008: Ahupua'a—Proceedings of the world environmental and water resources congress 2008, vol 316, pp 1–11. Retrieved from https://doi.org/10.1061/40976(316)285

Bharath A, Manjunatha M, Tangadagi RB, Preethi S (2020a) J Green Eng (JGE). Environmental Assessment for Rainwater Harvesting at GITAM Campus 10(4):1776–1785

Bharath A, Preethi S, Manjunatha M, Tangadagi RB (2020b) Prediction of temperature data for Ghataprabha Sub- 26:140–144

Derdous O, Djemili L, Bouchehed H, Tachi SE (2015) A GIS based approach for the prediction of the dam break flood hazard—A case study of Zardezas reservoir 'skikda, Algeria'. J Water Land Dev 27(1):15–20. Retrieved from https://doi.org/10.1515/jwld-2015-0020

Froehlich DC (2008) And their uncertainties. Environ Prot 134:1708–1721

Khosravi K, Rostaminejad M, Cooper JR, Mao L, Melesse AM (2019) Dam break analysis and flood inundation mapping: The case study of Sefid-Roud Dam, Iran. In: Extreme hydrology and climate variability: monitoring, modelling, adaptation and mitigation. Elsevier Inc. Retrieved from https://doi.org/10.1016/B978-0-12-815998-9.00031-2

Kulkarni SR, Ukarande SK, Jagtap S (2016) Dam break analysis—a case study. Int J Eng Res 5(Special 1):207–209

Mohamed MMA (2018) Overtopping breach peak outflow approximation of embankment dam by using Monte Carlo method. Beni-Suef Univ J Basic Appl Sci 7(4):724–732. Retrieved from https://doi.org/10.1016/j.bjbas.2018.10.002

USACE (2016) HEC-RAS river analysis system hydraulic reference manual version 5.0. Hydrologic Engineering Center, p 547

Wang B, Zhang T, Zhou Q, Wu C, Chen YL, Wu P (2015) A case study of the Tangjiashan landslide dam-break. J Hydrodyn 27(2):223–233. Retrieved from https://doi.org/10.1016/S1001-6058(15)60476-0

Xiong Y (2011) A dam break analysis using HEC-RAS. J Water Resour Prot 3(06):370–379. Retrieved from https://doi.org/10.4236/jwarp.2011.36047

Zhang L, Xu Y, Jia JS (2009) Analysis of earth dam failures: a database approach. Georisk 3(3):184–189. Retrieved from https://doi.org/10.1080/17499510902831759

Chapter 12
Suitability and Performance of Present Irrigation System in Kokernag, Jammu and Kashmir

Dar Himayoun, Roshni Thendiyath, and Jahangeer Saleem

Abstract This study deals with the suitability and performance of the irrigation system in the Kokernag, a hilly and mountainous area in the state of Jammu and Kashmir. Based on the variation in the soil texture, soil depth and output of paddy fields, the whole study area was divided into two zones, namely zone A, covering the middle portion of the Kokernag valley constituting the main paddy fields and zone B, covering the area around the nallah Bringi. The suitability of this irrigation system was assessed using parametric approach. This parametric evaluation was based on soil texture, soil depth, slope and drainage of that area. These factors were rated (0–100) based on their characteristics and importance for surface irrigation. Then, capability index was developed for each zone using the rating of above mentioned factors. On comparing this capability index with various suitability classes, it was found that the zone A was highly suitable, whereas zone B was marginally suitable for surface irrigation system. Further, the suitability analysis was carried on the basis of the availability of water and socio-economic background of that area. It was concluded that the existing system of irrigation is suitable in both the zones. The performance was evolved using physical and financial performance indicators in both the zones. Based on these performance indicators, the irrigation system in zone A was performing better than that of zone B.

Keywords Surface irrigation · Nallah Bringi · Capability index · Suitability and performance · Parametric evaluation

D. Himayoun (✉) · R. Thendiyath
Department of Civil Engineering, National Institute of Technology Patna, Patna, India

R. Thendiyath
e-mail: roshni@nitp.ac.in

J. Saleem
Department of Civil Engineering, National Institute of Technology Srinagar, Srinagar, India

R. Jha et al. (eds.), *Water Resources Management and Reservoir Operation*, Water Science and Technology Library 107,
https://doi.org/10.1007/978-3-030-79400-2_12

12.1 Introduction

Plants like human beings require air and water for their survival. Water is naturally supplied to these plants by nature through direct rain or through flood water. But there are some places where the distribution of rain and flood water is not uniform, in that case we need some artificial measures to feed these crops and this technique of applying water to the crops artificially is called irrigation.

A suitable method of irrigation is necessary for efficient use of water, as irrigation is considered as a vehicle for economic and agricultural development. It also plays a vital role in employment creation, improved nutrition, income generation, food security and raising the living standard of farmers. There is scarcity of fresh water resources in the world and irrigation is the major user of the fresh water supplies. With limited resources and increasing competition for these resources, there should be optimal utilisation of irrigation water (Elshaikh et al. 2018). How best an irrigation system performs its designated function, may depend on various variables such as climatic conditions, price, infrastructure design, socio-economic settings, availability of inputs and more important the management of the irrigation system (Ahmad et al. 2017; Elshaikh et al. 2018; Fang et al. 2018; Ghahroodi et al. 2015; Moreno et al. 2016; Salahou et al. 2018).

Kashmir valley has vast perennial fresh water sources. These resources have been utilised by mankind right from primitive time for various purposes. For irrigation purposes the sources of water supply are springs, nallahs, ground water, river tributaries and local streams. Due to availability of its vast fertile land, Kokernag has a very good potential of agricultural production. Amongst all the tehsils of district Anantnag in Jammu and Kashmir, Kokernag has the highest agricultural production. There are various crops grown in this area in kharief season like paddy, maize, pulses, fodder and other vegetables. Amongst them paddy is the main crop for which irrigation is needed. The irrigation employed there is direct irrigation method. In order to take care of diurnal variations in direct method of irrigation, some pondage is created by constructing a weir or barrage across the river. This helps in rising the water level in the river, then water is directed into the off taking canal which flows under gravity. The method of application of irrigation water is contour basin type. Since for paddy crop, ponding of water is needed up to around five centimetre from ground surface, which on sloping land is very difficult that is why the fields in the Kokernag area are contour basin type.

The present irrigation process in this area is surface irrigation gravity flow type. This method involves more losses and hence is less efficient as compared to other irrigation method like drip irrigation or sprinkler irrigation. In order to increase the output of its fertile lands, there is considerable need to increase the efficiency of its network of irrigation canals. So that this area comes to the position of its self-sufficiency.

Evaluation of suitability of surface irrigation on the basis of parametric approach depends on topographical characteristics like slope, physical properties of soil and drainage properties of that area. This suitability can also be discussed on the basis of socio-economic background and water availability in that area (Alonso et al. 2019; Antonio et al. 2018; Dang et al. 2018; Morris et al. 2015; Repository 2005; Usman et al. 2015). For proper use of water through irrigation process, it is also important to assess the performance of that irrigation scheme. Performance evaluation can be done using physical performance indicators and financial performance indicators. The aim of this study is to evaluate suitability and performance of irrigation system in Kokernag using parametric approach and performance indicators, respectively. Evaluation of suitability and performance of irrigation system is needed to achieve an increase in overall efficiency and to improve water management practises. This study contributes to the available literature by presenting the aggregated diagnosis of suitability and performance of present irrigation system to the decision-makers. This is useful to make them aware of the overall performance and steps to be taken on achieving sustainable balance between irrigation management and environmental preservation for an irrigation scheme.

12.2 Methodology

The methodology for evaluating suitability of this surface irrigation system in the study area was discussed on the basis of three approaches.

12.2.1 Parametric Approach

In the parametric evaluation approach, land is evaluated on the basis of numerical indexes (Albaji et al. 2001; Dengiz et al. 2006). This evaluation approach is based on physical and chemical properties of soil. In this approach, soil is classified by rating the land characteristic from 0 to 100 and comparing with the tables of soil requirement. These ratings are used to calculate the multiplicative index of the soil. Another classification of land resources is based on land capability. This classification depends on climatic parameters, soil and topography and does not take into account socio-economic conditions and yield. In this classification, negative features of land are highlighted, whilst allotting capability classes to the lands of different types.

12.2.2 On the Basis of Socio-Economic Background

Socio economic status of a society is the combined measure of both economic and sociological position in relation to the other society based on the income, education and occupation. Here in this area, most of the families are of lower class and agriculture is their main source of income. Most of the people are unskilled and unemployed. The idea of taking this factor into account was that the surface irrigation scheme require large number of labours and they need not be skilled. To some extent, this irrigation system solves the unemployment problem there, since there are large number of unskilled labours available. Also the other systems like-drip irrigation system, sprinkler irrigation system and sub-surface irrigation system require high-initial cost, skilled labour, more electricity and proper management and maintenance, which in majority of cases is not available in this area.

12.2.3 On the Basis of Usage and Availability of Water

Type of crop affects the choice of selecting a particular irrigation method since different crops require different amount of water for their growth. Paddy and sugarcane require a large quantity of water as compared to other crops. For paddy, crop ponding of water has to done in the field continuously for about seventy to eighty days, this requirement is cheaply fulfilled by surface irrigation if there is abundance of water available. Here in the present case, the availability of water is much more than the total requirement there. Since in surface irrigation, there are more losses as compared to other methods of irrigation that means choice of selecting a particular type of irrigation also depends upon the water availability. If there is plenty of water available then we can go for surface irrigation and if there scarcity of water, then we have to go for other methods of irrigation that are more conservative.

12.3 Suitability and Performance Indicators

12.3.1 Capability Index

The capability index (CI) for surface irrigation was calculated on the basis of Eq. (12.1) given below

$$\mathrm{CI} = \frac{A \times B}{100} \times \frac{C}{100} \times \frac{D}{100} \tag{12.1}$$

where 'A' is the soil texture rating, 'B' is the soil depth rating, 'C' is the slope rating and 'D' is the drainage rating. The ranges of capability index and the corresponding suitability classes are presented in Table 12.1.

Table 12.1 Suitability classes for various capability indices

Capability index	Class	Suitability class	Symbol
>80	1	Highly suitable	S1
60–80	2	Moderately suitable	S2
45–60	3	Marginally suitable	S3
30–45	4	Currently not suitable	N1
<30	5	Permanently not suitable	N2

12.3.2 Relative Water Supply (RWS)

RWS is a measure of adequacy as well as seasonal timeliness of an irrigation system. It is a dimensionless parameter and is defined as the ratio of total water supply to the total demand at scheme level (Eq. 12.2). The numerator is the irrigation water supply and the denominator consists of non-beneficial evapotranspiration (ET), losses to drains, consumptive use and net flow towards groundwater.

$$\text{RWS} = \frac{(\text{Irrigation water})}{(\text{ET} + \text{Seepage} + \text{Percolation})} \tag{12.2}$$

There are three ranges for RWS—low RWS (<1), medium RWS (1–2) and high RWS (>2). In low RWS, the water supply is below actual requirements, requires strict control and higher degrees of management. In medium RWS, a variety of operational strategies can be successfully employed, moderate amounts of management effort are needed and high RWS has more likely the farmer controlled demand schedules and greatest flexibility in operation.

12.3.3 Benefit—Cost Ratio (BCR)

The financial performance of irrigation system is evaluated on the basis of benefit—cost ratio (Eq. 12.3). This indicator gives the relation between the total input and total output in terms rupees. The ratio should be greater than one for all irrigation systems. The benefit—cost ratio was worked out for the irrigation system in both the zones and it was greater than one for both the zones, zone '*A*' having higher valve.

$$\text{Benifit} - \text{Cost ratio} = \frac{\text{Output cost}}{\text{Input cost}} \tag{12.3}$$

12.4 Description of Study Area

12.4.1 Location

Kokernag is a sub-district town and a notified area in Breng Valley. This area lies in district Anantnag in state of Jammu & Kashmir. Kokernag is located towards southern part of the Kashmir valley in the foothills of famous Pir-Panjal mountain range and is at a distance of 25 km from Anantnag and about 80 km from Srinagar city. Kokernag area is the catchment area of nallah 'Bringi'. Bringi is one of the tributaries of the river Jhelum, hence this catchment is considered as one of the upland catchments of Jhelum river. Bringi catchment lies between 33.20° N and 33.45° N latitude and 75.10° E and 75.30° E longitude covering a total area of 595 km^2. The elevation varies from 1650 to 4000 m above mean sea level. Figure 12.1, shows the location map of study area.

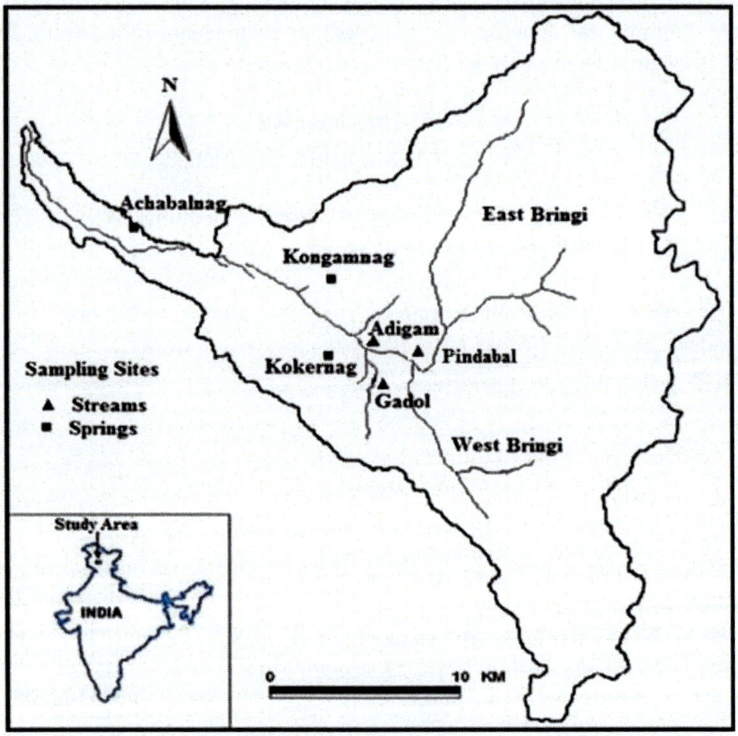

Fig. 12.1 Location map of the study area

12.4.2 Topography

Kokernag is hilly and mountainous towards the southeast and southwest with broad inter-mountain valley on the northern side. The altitude of the hill ranges up to 3000 m above mean sea level. The valley in the central part has mildly undulating topography with its elevation ranging from 1800 to 2000 m above mean sea level. The soils are fertile mostly silty and sandy silt in texture and the main crop grown is paddy. The land features of the study area are presented in Table 12.2.

12.4.3 Precipitation Characteristics

The rainfall characteristics in Jammu and Kashmir are different from rest of India. In India, most of the rainfall (about 75%) occurs in monsoon season, that is, from June to October. In Jammu and Kashmir most of the rainfall occurs in winter season and spring season, that is, from January to April. Table 12.3, shows the 12 year averaged monthly rainfall at station Kokernag.

Table 12.2 Land features of the study area

Irrigated area (ha)	Unirrigated area (ha)	Orchards (ha)	Grazing area (ha)	Total (ha)
1627	2015	843	1114	5609

Table 12.3 Average monthly rainfall at station Kokernag

Month	Rainfall in mm
January	64
February	125
March	113
April	106
May	92
June	71
July	85
August	69
September	66
October	25
November	26
December	54
Total	895

12.4.4 *Irrigation Details*

12.4.4.1 Source of Irrigation

Nallah Bringi is the main source of irrigation in Kokernag. It is located in district Anantnag of Jammu and Kashmir and starts from Wandevalgam village which is located at a distance of 30 km from Anantnag main town and ends near Larkipora village which is located in the out skirts of Anantnag. Bringi catches snowmelt water and becomes active during summer season when there is high-intensity rainfall in the upper catchments. The average discharge in the Bringi nallah varies from 12 to 75 cumecs. This is sufficient to meet the requirement of that area. The weekly discharge variation in Bringi nallah is presented in Fig. 12.2.

12.4.4.2 Type of Irrigation

The irrigation scheme adopted in Kokernag is direct irrigation method flowing under gravity. The method of application of irrigation water is contour basin type. Since for paddy crop, ponding of water is needed up to around five centimetre from ground surface which on sloping land is very difficult that is why the fields there are contour basin type and hence the need of contour basin irrigation.

12.4.4.3 Types of Crops Grown

There are various crops grown in this area in kharief season like—paddy, maize, pulses, fodder and other vegetables. Among them paddy is the main crop for which irrigation is done. Table 12.4, shows the area covered by the crops grown in the study area.

The total irrigated area in Kokernag is around 1627 ha. Out of this, 1499 hectares of area is under paddy crop, whilst remaining is under the rest of crops grown there.

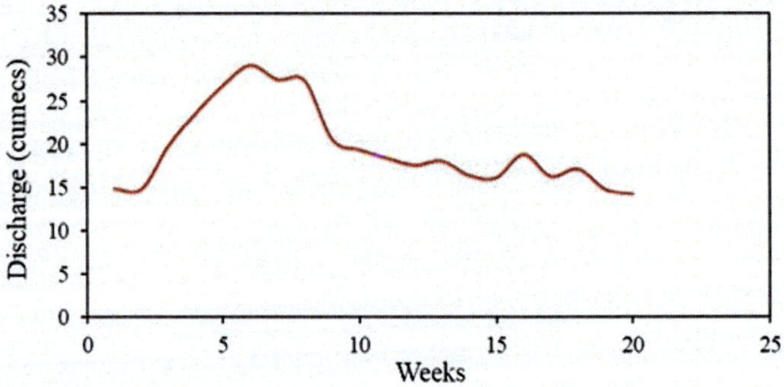

Fig. 12.2 Weekly variation in Bringi discharge

Table 12.4 Crops grown in the study area

Crop	Paddy	Pulses	Maize	Vegetables	Fodder
Area in ha	1499	74	1855	129	56

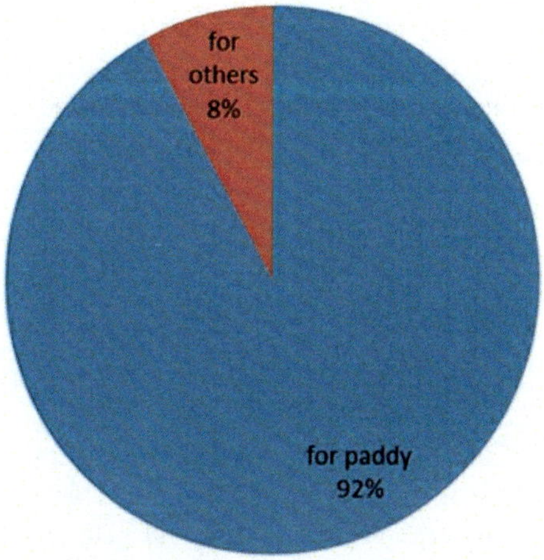

Fig. 12.3 Irrigation requirement in Kharief season

Figure 12.3, shows the total irrigation requirement in kharief season in Kokernag area.

The water requirement for paddy crop starts early in the month April for the development of nurseries for paddy crop, but that is in small quantities. It is about one-tenth of that required for normal paddy crop after transplantation. After transplantation, the water requirement rises and is at peak during the month of June and July. Then, water requirement starts decreasing and reaches zero at the mid of September.

12.5 Data Collection

The data was collected from the various departments like agricultural department division Achabal Anantnag, irrigation department division Dooru Anantnag, watershed management department Anantnag and Indian meteorological department Srinagar. Due to non-availability of some data in these departments, some data was collected locally and some on interview with different farmers. The different types of data collected are; soil data, topographical data, crop production data, water availability and requirement data and some data related to socio- economical background.

12.5.1 Soil Data

This includes the information regarding the soil texture and soil depth. Different sites were chosen for this data requirement. In the middle areas of the Kokernag, the soil depth varies from 100 to 200 cm, but in the area around the nallah Bringi, it was ranging from 25 to 50 cm with an average of 40 cm. Beyond this soil cover depth, there is mixture of gravel, sand, silt and some stones. For determination of soil texture, two different soil samples were taken. One from Bringi region and another from the main paddy field region, i.e. from middle of the Kokernag area. From sieve analysis and hydrometric analysis, particle size distribution curves of both the soil samples were drawn as shown in Figs. 12.4 and 12.5, respectively.

12.5.2 Water Availability and Water Requirement Data

The discharge data of the Bringi region and the water requirement for paddy crop were obtained from the irrigation department Vailoo. Table 12.5, presents the average weekly variation in discharge of Bringi at Wayul gauge, Table 12.6, presents the monthly requirement of water for paddy Crop in middle fields of Kokernag and Table 12.7, presents the monthly requirement of water for paddy crop in area around nallah Bringi Kokernag.

Now from the above data, we see that the middle portion of the valley considerably differ from the area around the nallah Bringi keeping in view the soil texture, soil depth and water requirement in the fields. It is because of these variations that the whole study area was divided into two zones: Zone A—this includes the area covering

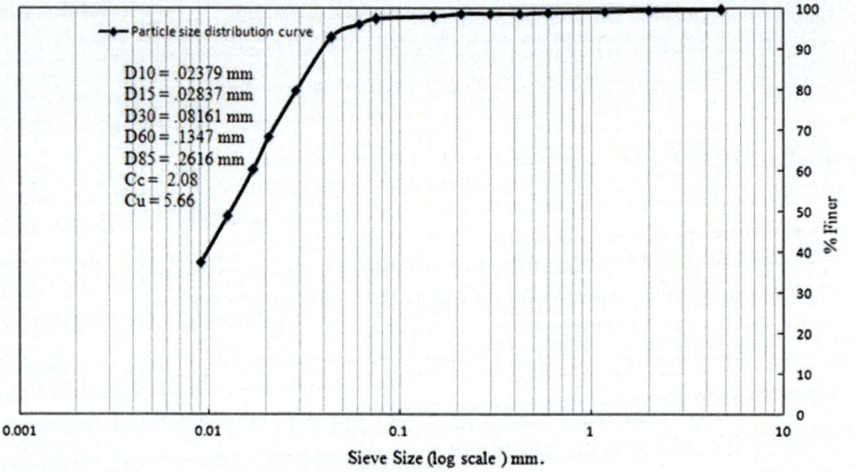

Fig. 12.4 Particle size distribution curve for soils in middle fields of Kokernag

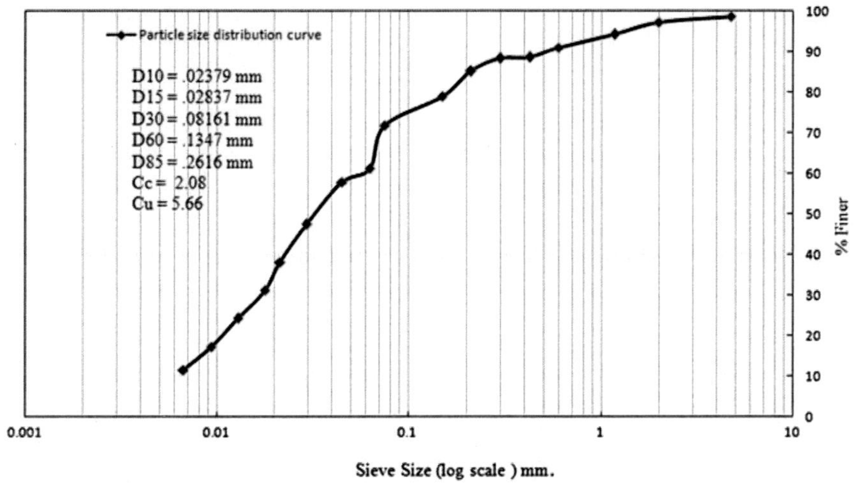

Fig. 12.5 Particle size distribution curve for soils around Bringi zone

Table 12.5 Average weekly variation in discharge of Bringi at Wayul gauge

Month	Discharge (m^3/s) (1st week)	Discharge (m^3/s) 2nd week	Discharge (m^3/s) 3rd week	Discharge (m^3/s) 4th week
April	14.82	14.79	19.78	23.52
May	26.7	29.07	27.43	27.29
June	20.51	19.3	18.36	17.54
July	18.02	16.38	16.09	18.76
August	16.24	17.09	14.71	14.16
September	13.43	12.1	–	–

Table 12.6 Showing the monthly requirement of water for paddy crop in middle fields of Kokernag

Month	Delta for paddy in mm
January	–
February	–
March	–
April	–
May	200
June	300
July	300
August	225
September	150
October	–
November	–
December	–
Total	1175

Table 12.7 Showing the monthly requirement of water for paddy crop in area around nallah Bringi Kokernag

Month	Delta for paddy in mm
January	–
February	–
March	–
April	–
May	280
June	420
July	420
August	315
September	210
October	–
November	–
December	–
Total	1645

the middle portion of the Kokernag valley that constitutes the main paddy fields. This area is about the 90% (1350 ha) of the total paddy field area. And Zone B—this includes the area around the Bringi stream constituting the Bringi paddy fields. This covers an area of about 10% (150 ha) of the paddy field area.

12.6 Results and Discussions

12.6.1 Suitability

Based on the variation in soil texture, soil depth and output from the fields, the whole study area was divided into zones. Zone A consists of the main paddy fields which lie in the middle of the Kokernag valley. The other zone was the area which lie around the Bringi Nallah named as zone B. Zone A constitute about 1350 ha of paddy field area where as zone B constitute an area of around 150 ha of paddy land. In order to determine the soil texture in these zones, soil samples were taken from both the zones and particle size distribution curves of both the samples were drawn using sieve analysis and hydrometric analysis. The soil from zone A was found to be silty soil and the soil from the zone B was found to be sandy silt soil. Also the permeability tests were done on the soil samples, taken from zone A and zone B, at optimum moisture content. This could not represent the actual field conditions, but it gave us relative permeability valves. The permeability in zone B was found to be twice that of zone A. For calculating the soil depth in both the zones, some pits were dug up in both the zones. In zone A, the average soil cover depth was found to be more than 100 cm and in zone B it was around 40 cm with minimum thickness of about 25 cm.

Table 12.8 Suitability and performance indices

	Soil type	
	Zone A	Zone B
Performance indices		
RWS	1.6	1.71
BCR	1.83	1.4
Suitability index		
CI	85.5	46

The crop in both the zones, for which suitability of irrigation system is to be evolved, was chosen paddy crop and both the paddy field zones get irrigation from the same source—'Bringi stream'. Since the paddy fields require large quantities of water and increases the chances of water logging but in both the sites there was no area where these conditions could occur. It was because of natural slope present there which could drain off these fields quickly and easily.

The suitability of this surface irrigation system was evolved using parametric approach (Sys et al. 1991). Capability index for both the was calculated and based on this index the land in zone A was found to be highly suitable for the surface irrigation with the CI value of 85.5, whereas the land in zone B was found to be marginally suitable for the same with the value of 46. Moreover, the type of irrigation employed also depends on the factors like—capital cost, operating cost, ease of use and efficiency potential. The surface irrigation system there reduces the operating costs of irrigation—principally electricity, which was cited as a major cause of failure in that area. The surface irrigation system is the least expensive to implement, but does require more labour than the other systems. However, with the high unemployment levels in the area, this had become a positive factor for that economy. Also other systems of irrigation requires more skilled labours and the availability of these was difficult in that area that is why the present irrigation system has become easier to operate. A comparison was made between the crop water requirement and availability of water there. It was found that the water available for irrigation during peak demands in both the zones was much more than the requirement. The suitability and performance indices of the irrigation system are presented in Table 12.8.

12.6.2 Performance

The performance of this irrigation system in these two zones was evaluated using the physical performance indicators and financial performance indicator. Physical performance indication was based on the relative water supply and irrigation water-yield. The financial performance indication was based on the benefit–cost ratio.

In zone A, the relative water supply was around 1.6, this means the water supplied there is higher than the water requirement. The reason behind this are various losses like-seepage, application losses and tail water losses. Based on the irrigation water-yield indication, it was estimated that 1 kg of rice at an average requires 4500 L of water which is very high as compared to the average requirement of about 2000–3000 L based on the International Water Management Institute standard. For financial performance, benefit–cost ratio was calculated, and its value was found around 1.83. This means that the net return from the system is greater than the initial investment.

In zone B, the relative water supply was around 1.71 which is higher as compared to that of zone A. The reason behind this is the presence of sand content in that soil, which increases seepage and other losses. Based on the irrigation water yield, it was estimated that 1 kg of rice requires about 9800 L of water which when compared with that of zone A is very large. The possible reason for this high-water use has been already discussed, that is, the presence of sandy soil there and less soil depth. This high-water use increases the percolation losses and other seepage losses, and decreases the output of this area. The financial performance in this zone, based on benefit cost ratio, was about 1.41 that means output from this area is very low as compared to that of zone A. The possible reason for this low output may be the thin soil cover. Since the root zone depth is greater than the soil depth and the presence of gravel and stones below the soil cover limit the potential growth of the paddy root which ultimately affected the output of this area.

12.7 Conclusions

Although based on parametric approach, the land in zone A was highly suitable whereas the land in zone B was marginally suitable for surface irrigation, but when suitability was evaluated on the basis of socio-economic background and availability of water in the study area, it was found that surface irrigation system is suitable in both the zones. Also area under zone A (90%) is much more than zone B (10%) that means the additional water losses that occur in zone B are negligible as compared to the total water use in both the zones. Hence from the above study it was concluded that the existing system of irrigation is suitable in the study area. Now for performance, we have observed that the performance of irrigation system in zone A is better than that of zone B because of low-percolation losses and higher yield in zone A. Also the performance on the basis of relative water supply and irrigation water-yield is very low in both the zones, but it is on the basis of benefit–cost ratio that makes the performance acceptable in both the zones.

References

Ahmad E, Brar AS, Mishra SK, Singh KB (2017) Field crops research simulating response of wheat to timing and depth of irrigation water in drip irrigation system using CERES-wheat model. Field Crops Res 214(04):149–63. https://doi.org/10.1016/j.fcr.2017.09.010

Albaji M, Boroomand Nasab S, Hemadi J (2001) Comparison of different irrigation methods based on the parametric evaluation approach in West North Ahwaz Plain. Agric Water Manage 208(05):193–203. https://doi.org/10.1016/j.agwat.2018.05.018

Alonso A, Feltz N, Gaspart F, Sbaa M, Vanclooster M (2019) Comparative assessment of irrigation systems' performance : case study in the Tri Ff a agricultural district, NE Morocco. Agric Water Manage 212(01):338–48. https://doi.org/10.1016/j.agwat.2018.08.033

Antonio D, Nicotra A, Mateos L, Marcello S (2018) Improvement of the irrigation performance in water users associations integrating data envelopment analysis and multi-regression models. Agric Water Manage 205(05):38–49. https://doi.org/10.1016/j.agwat.2018.04.032

Dang T, Pedroso R, Laux P, Kunstmann H (2018) Development of an integrated hydrological-irrigation optimization modeling system for a typical rice irrigation scheme in Central Vietnam. Agric Water Manage 208(05):193–203. https://doi.org/10.1016/j.agwat.2018.05.018

Dengiz O (2006) Comparison of different irrigation methods based on the parametric evaluation approach. Agric Water Manage 30(3):21–29

Elshaikh AE, Jiao X, Yang SH (2018) Performance evaluation of irrigation projects: theories, methods, and techniques. Agric Water Manage 203(01):87–96. https://doi.org/10.1016/j.agwat.2018.02.034

Fang Q, Zhang X, Shao L, Chen S, Sun H (2018) Assessing the performance of different irrigation systems on winter wheat under limited water supply. Agric Water Manage 196(02):133–43. https://doi.org/10.1016/j.agwat.2017.11.005

Ghahroodi EM, Noory H, Liaghat AM (2015) Performance evaluation study and hydrologic and productive analysis of irrigation systems at the Qazvin Irrigation Network (Iran). Agric Water Manage 148(06):189–95. https://doi.org/10.1016/j.agwat.2014.10.003

Moreno MA, Castillo A, Montero J, Tarjuelo JM (2016) ScienceDirect optimisation of the design of pressurised irrigation systems for irregular shaped plots. Biosyst Eng 151(05):361–73. https://doi.org/10.1016/j.biosystemseng.2016.10.005

Morris MR, Hussain A, Gillies MH, Halloran NJO (2015) Inflow rate and border irrigation performance. Agric Water Manage 155(04):76–86. https://doi.org/10.1016/j.agwat.2015.03.017

Repository, Institutional (2005) Institutional repository performance assessment of irrigation water management of heterogeneous irrigation schemes: 1. A framework for evaluation this item was submitted to Loughborough's Institutional Repository by the Author and Is Made Available Un

Salahou MK, Jiao X, Lü H (2018) Border irrigation performance with distance-based cut-off. Agric Water Manage 201(01):27–37. Retrieved (https://doi.org/10.1016/j.agwat.2018.01.014).

Usman M, Liedl R, Awan UK (2015) Spatio-temporal estimation of consumptive water use for assessment of irrigation system performance and management of water resources in irrigated Indus Basin, Pakistan. J Hydrol 51(05):361–73. https://doi.org/10.1016/j.jhydrol.2015.03.031

Chapter 13
Linking of Sediment Yield Pattern with Rainfall and Land-Use Land-Cover Changes Within Burhanpur Sub-catchment, India

S. R. Resmi, P. L. Patel, and P. V. Timbadiya

Abstract In present study, the temporal changes in the suspended sediment concentration and sediment yield pattern of Burhanpur sub-catchment, which is a part of Upper Tapi basin, India, are investigated. The variation of sediment yield during three decades, viz. 1980–1989, 1990–1999 and 2000–2010 is correlated with the changes in the land-use land-cover and climatic pattern within the sub-catchment. The changes in the pattern of monsoon rainfall intensity were investigated using modified Mann–Kendall (MMK) test while land-use land-cover within the sub-catchment was investigated by classifying the decadal images of the sub-catchment. The analysis of observed data revealed that sediment concentration is very high, even for small-to-moderate flows during the onset on monsoon, due to availability of unconsolidated sediments after a long dry period of October–May months. The monsoon rainfall intensity has shown decreasing trend over the sub-catchment during the period 1980–2010. Also, agricultural areas are found to be increasing with time at the expense of forest cover, fallow and scrub lands. The waterbodies, in the form of minor storage structures, are increased during the period 1990–2010. The excessive trapping of sediments within the sub-catchment due to the development of minor storage structures (waterbodies) was responsible for reduction of sediment yield within the Burhanpur sub-catchment.

Keywords Sediment yield · Land-use land-cover · Modified Mann–Kendal test · Monsoonal rainfall intensity · Waterbodies

---

S. R. Resmi · P. L. Patel (✉) · P. V. Timbadiya
Sardar Vallabhbhai National Institute of Technology Surat, Surat 395007, India

R. Jha et al. (eds.), *Water Resources Management and Reservoir Operation*,
Water Science and Technology Library 107,
https://doi.org/10.1007/978-3-030-79400-2_13

13.1 Introduction

The geomorphic characteristics of a watershed affect the energy fluxes, mass movement and dispersion of water and sediment within the watershed. Soil erosion induces detachment and transport of sediments and is treated as one of the dominant geomorphic processes on earth's land surface (Toy et al. 2002). The sediment is detached from the soil surface by impacts of raindrops and by the shearing force of surface runoff (Jain and Kothyari 2000). In case the sediments available for transport are greater than transport capacity of flow, the sediment gets accumulated on the land surface (Trimble 1975; Toy et al. 2002). The amount of sediment that arrives at the outlet of a catchment is termed as sediment yield (Glymph 1954; Parsons et al. 2006; Fryirs 2013; Vanmaercke et al. 2014). This sediment yield within a catchment is determined by its environmental conditions, such as climate, soil, topography, land use, other human disturbances, which can affect the sediment supply, transport, storage, residence time and the connectivity of sediment sources to the catchment outlet (Mingguo et al. 2008; Fryirs 2013; Shi et al. 2013; Yan et al. 2013). The quantification of sediment yield and its determinant factors is of utmost importance for sustainable management of catchments (Parsons et al. 2006). Sediment yields are sensitive to the factors like reservoir construction, land clearance and land-use changes as well as other forms of land disturbances, such as mining activity, soil and water conservation measures, sediment control programmes and climate change (Walling and Fang 2003). Human activities and climate change have profound impacts on geomorphic characteristics of catchments (Fryirs 2013). Human-induced activities, such as construction of dams and reservoirs, decrease the sediment yield, whereas clearance of natural vegetation and mining activities is responsible for reduction in the flow resistance over the soil surfaces, thereby increasing the sediment yield. Land cover is one of the major factors that reduces soil erosion due to improvement in soil structure, increase in the surface roughness and infiltration rate, trapping the sediments and breaking the landscape connectivity (Burylo et al. 2007; Hudek et al. 2010, Ouyang et al. 2010; Preti et al. 2011; Lü et al. 2012; Rey and Burylo 2014). Human land-use activities can fundamentally change the hydrogeomorphology of rivers (James and Lecce 2013). Such changes in land use can have significant impact on the runoff, sediment delivery, water quality, aquatic habitat and channel morphology. Human impacts on rivers can be classified as 'direct changes', which occur within channels and the riparian vegetation and 'indirect changes', which occur in the form of agricultural expansion, urbanization, mining and changes in the vegetation. Apart from anthropogenic changes within the catchment, the climatic factors like changes in rainfall pattern are also responsible for changing the sediment yield pattern at the outlet of the catchment. The present study is an attempt to quantify the changes in suspended sediment concentration and sediment yield at Burhanpur stream gauging station over a period of 30 years, i.e. 1980–2010. The study has been carried out to quantify changes on decadal basis, i.e. during 1980–1989, 1990–1999 and 2000–2010. The variations in sediment yield pattern due to changes in monsoonal rainfall intensity and land-use land-cover are linked together for Upper Tapi sub-basin (Burhanpur sub-catchment).

The study contributes significantly in understanding of the effects of climatic and land-use land-cover changes on sediment yield pattern within the sub-catchment.

13.2 Study Area

The Tapi river is the second largest west flowing river in India, after Narmada river. Originating from the Tapi talab near Multai in Betul district of Madhya Pradesh, the river traverses around 784 km till it joins the Arabian Sea at the Gulf of Cambay in Gujarat. The study area of interest in present study, Burhanpur sub-catchment of the Upper Tapi basin, is located between latitudes 21° 00′ 00″ N to 22° 00′ 00″ N and longitudes 75° 45′ 00″ E to 78° 30′ 00″ E and drains an area of 10,612 km^2 to the Hatnur reservoir (Fig. 13.1). The Upper Tapi river, which has a length of around 350 km upto Hatnur dam, is perennial in nature and has two stream gauging stations, viz. Dedtalai and Burhanpur, located at CH: 236.34 km and CH: 307.89 km, respectively. The main Tapi river is monsoon dominated, with significant contribution from south-west monsoon season that starts in mid-June and subsides in the mid-October. The July and August months are the rainiest months of the year, which contributes to around 50% of the total rainfall. The monsoon period is followed by a

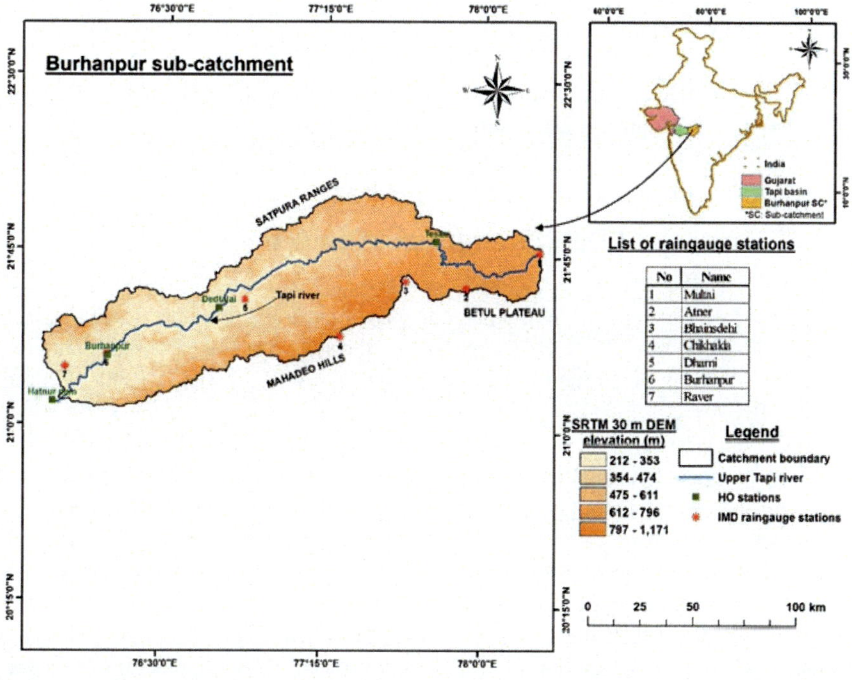

Fig. 13.1 Study area: Burhanpur sub-catchment

long dry season, which spans over 8 months (October–May). There are seven rain-gauge stations in the sub-catchment, which are operated and maintained by India Meteorological Department (IMD). The Upper Tapi river passes through a variety of topography, such as gullied ravenous terrains, hilly areas and alluvial plains. The Upper Tapi basin (Burhanpur sub-catchment) stationed at an average altitude of 487 m from the mean sea level (MSL) with an average gradient of 6.26°, encompasses main Tapi river of 6th order. The Upper Tapi river lies in close proximity to Satpura ranges (average distance = 38 km), accordingly, northern tributaries lie in steep sloped areas, thereby exhibiting flashy streamflow characteristics. The Burhanpur sub-catchment is predominantly occupied by Deccan trap basaltic lava formations, followed by quaternary alluvial deposits towards the downstream of Dedtalai stream gauging station. Also, rock formations like granite-gneiss, quartz mica and sandstone deposits are present in minor quantities, especially, in the north-eastern part of the sub-catchment near the Betul region. The Deccan trap basaltic lava formation, granite-gneiss deposits and sandstone traces are the main sources of coarser sediments, viz. cobbles and boulders, in the headwater region of the sub-catchment, whereas the quaternary alluvial deposits in Burhanpur and Purna sub-catchments contribute to the silt and clay sediments in the river, especially, during the monsoon period (Hire 1999).

13.3 Methodology

The steps followed in quantifying the temporal sediment yield pattern within the Burhanpur sub-catchment are described in Fig. 13.2.

13.4 Analysis of Data, Results and Discussions

The observed suspended sediment concentrations at Burhanpur stream gauging station during 1980–2010 were used to obtain the sediment load and sediment yield. The analysis was performed on a decadal basis, i.e. 1980–1989, 1990–1999 and 2000–2010.

Preliminary inspection of the data reveals that, more than 99% of the sediment load is transported during the monsoon months (June–October) at Burhanpur stream gauging station. Accordingly, in present study, the monsoonal period has been taken into consideration for further analyses. For approximately 92% of the time, the concentration of suspended sediments is below 4 g/L at Burhanpur stream gauging station. This points to the fact that there is no adequate source of unconsolidated sediments in the catchment area upstream of the station. The presence of rocky bed stratum, upstream of Burhanpur stream gauging station, is responsible for scarcity of fine-grained material in the stream flow.

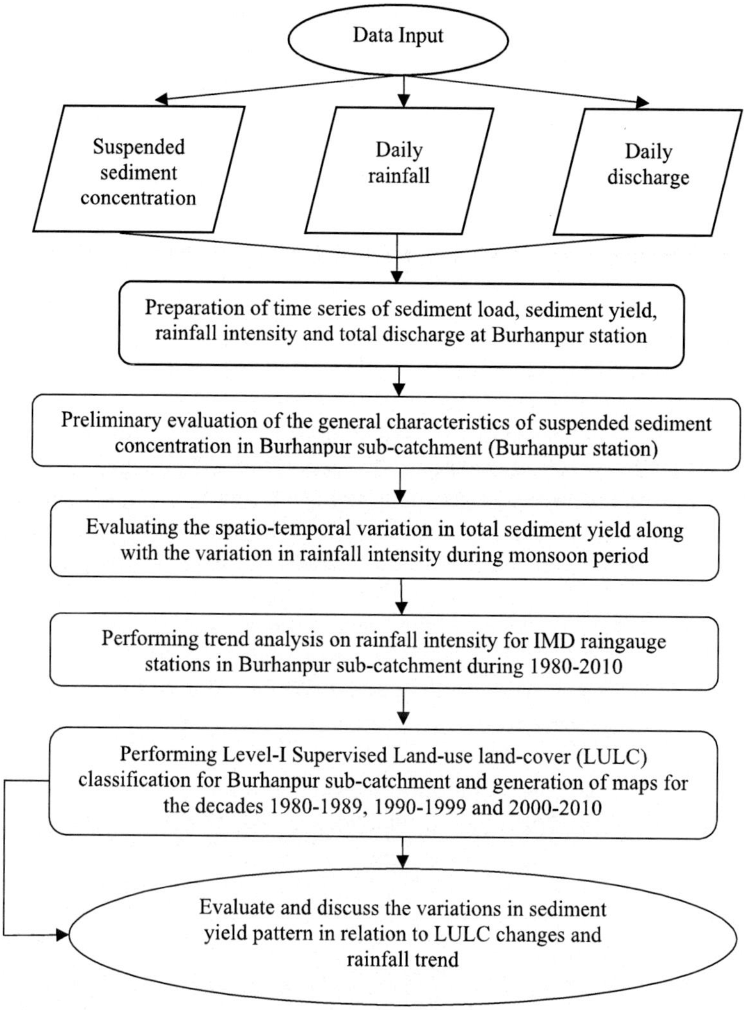

Fig. 13.2 Methodology adopted in present study

Table 13.1 summarizes the primary characteristics of the suspended sediment concentration and suspended sediment load at the outlet of Burhanpur sub-catchment of Upper Tapi river. The mean suspended sediment concentration values were observed to be 0.32 g/L during 1980–1989, 0.33 g/L during 1990–1999 and 0.24 g/L during 2000–2010.

The mean suspended sediment concentration values were observed to be decreasing over the decades. The maximum recorded suspended sediment concentration was 9.01 g/L during the period 1980–1989, which was in the year 1988, corresponding to a discharge of 4707 m^3/s and a rainfall of 85 mm (22/07/1988).

Table 13.1 Statistical characteristics of the suspended sediment at Burhanpur stream gauging station

Period	Suspended sediment concentration (g/L)			Suspended sediment load (t/day)			Average annual sediment yield (t/km^2/day)
	Mean	Min	Max	Mean	Min	Max (10^6)	
1980–1989	0.32	0.001	9.01	31,168	0	3.65	3.21
1990–1999	0.33	0	12.72	48,878	0	9.04	5.29
2000–2010	0.24	0	3.92	21,773	0	5.55	2.33

Similarly, during 1990–1999, a maximum observed suspended sediment concentration was 12.72 g/L on 16/07/1993, corresponding to a discharge of 8268 m^3/s and a rainfall of 126 mm over the sub-catchment. This indicates clearly that daily suspended sediment concentration is maximum during the onset of monsoon period, due to availability of unconsolidated sediment deposits, after a long dry span of eight months, i.e. from October to May in the sub-catchment. Even though the sediment sources are limited in Burhanpur sub-catchment due to its rocky topography in the upstream reaches, higher relief and steeper gradients are responsible for generating geomorphically effective flows for transporting the sediments downstream.

For Burhanpur sub-catchment, the average annual rainfall intensity (RI) during 1980–2010 was 20.3 mm/day for an average sediment yield of 3.57 t/km^2/day. The average sediment yield for the period 1980–1989 was 3.21 t/km^2/day, with the higher yields obtained in years 1983 (RI = 19.3 mm/day) and 1988 (RI = 22.1 mm/day). In the second decade, the average sediment yield observed was very high (5.29 t/km^2/day). The average sediment yield during 1990–1994 was 8.19 t/km^2/day, wherein RI during monsoon season ranges between 21 mm/day and 25.2 mm/day. On the other hand, during 2000–2010, for a rainfall intensity of 20.6 mm/day, average sediment was reported to be 2.33 t/km^2/day. While comparing Fig. 13.3a–c, it is seen that, in general, there is a reduction in the sediment yield from Burhanpur sub-catchment with effect from year 1994, with an average sediment yield of 2.40 t/km^2/day and 2.33 t/km^2/day during the period 1995–1999 and 2000–2010, respectively.

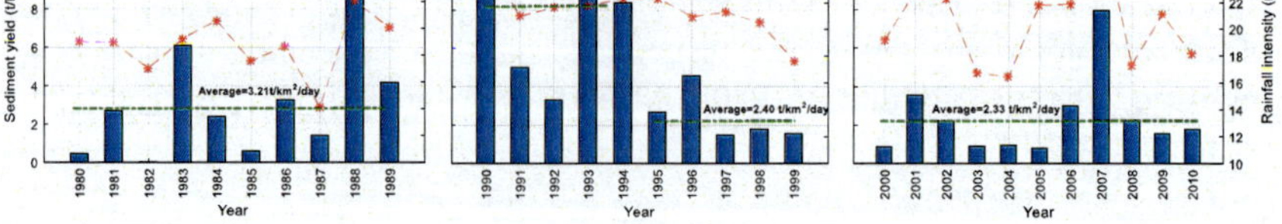

Fig. 13.3 Decadal variation in total sediment yield and rainfall intensity during monsoon period at the outlet of Burhanpur sub-catchment

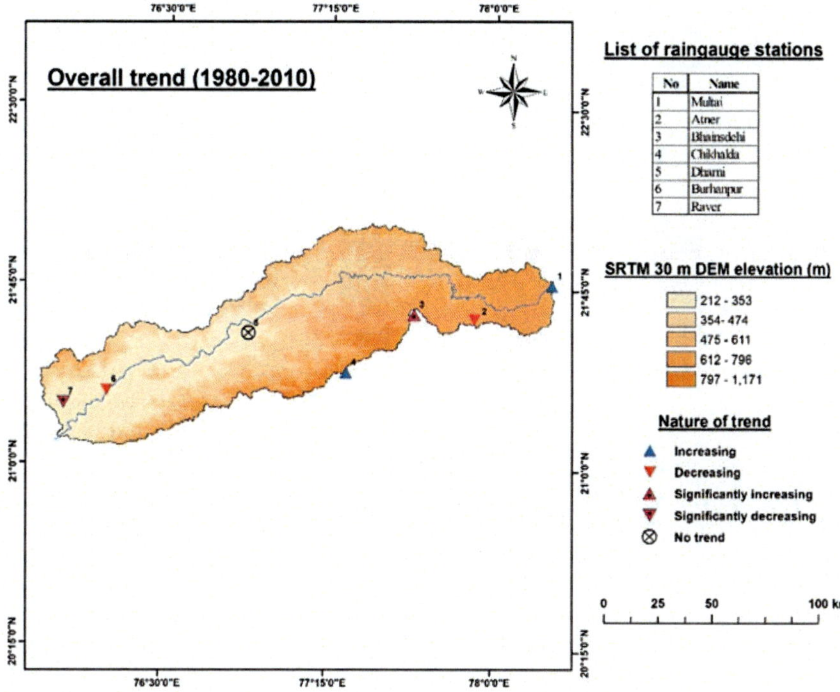

Fig. 13.4 Trend analysis results of rainfall intensity for Burhanpur sub-catchment

The pattern of total sediment yield was analysed while performing trend analysis of monsoonal rainfall intensity for seven raingauge stations in the sub-catchment using modified Mann–Kendall (MMK) test during the period 1980–2010. The results revealed that the monsoonal rainfall intensity in the sub-catchment exhibited decreasing trend during 1980–2010, especially, at stations Burhanpur and Raver, which fall in the alluvial reaches of the sub-catchment (Fig. 13.4).

The land-use land-cover statistics showed that there has been an increase in the agricultural area in the sub-catchment, at the expense of its forest cover (Fig. 13.5). Forest cover plays a major role in holding the soil in place, thereby preventing the soil erosion. The increase in agricultural activities like ploughing and tillage disturbs the soil surface, thereby increasing the sediment delivery to the river. However, this increasing effect is not observed in the Burhanpur sub-catchment. The spatio-temporal variations of waterbodies within the sub-catchment were also assessed and presented, along with the drainage network in Fig. 13.6. The analysis revealed that the waterbodies have expanded considerably during 1980–2010, even though they occupy only a small fraction of the total catchment area. The area of waterbodies existed within Burhanpur sub-catchment is 6.83 km^2, 17.72 km^2 and 20.05 km^2, respectively, during the decades 1980–1989, 1990–1999 and 2000–2010. This implies that construction of minor storage dams, irrigation weirs and minor check dams was responsible in trapping of sediments within the sub-catchment,

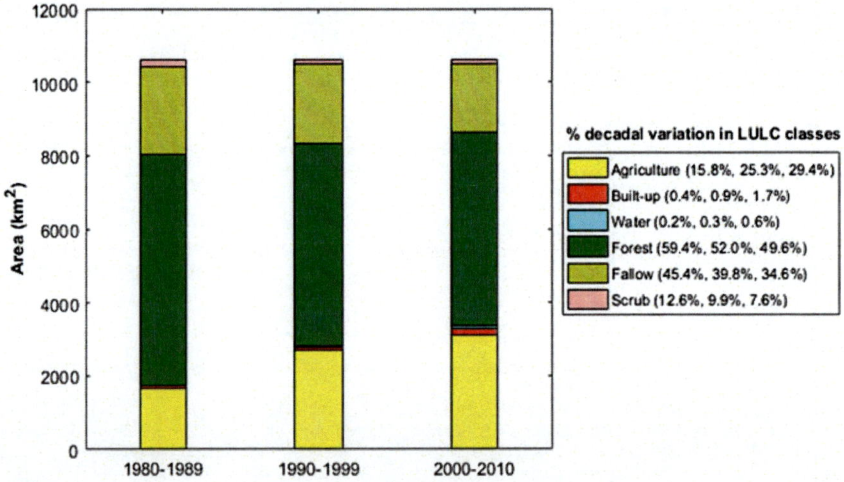

Fig. 13.5 Land-use land-cover pattern in Burhanpur sub-catchment

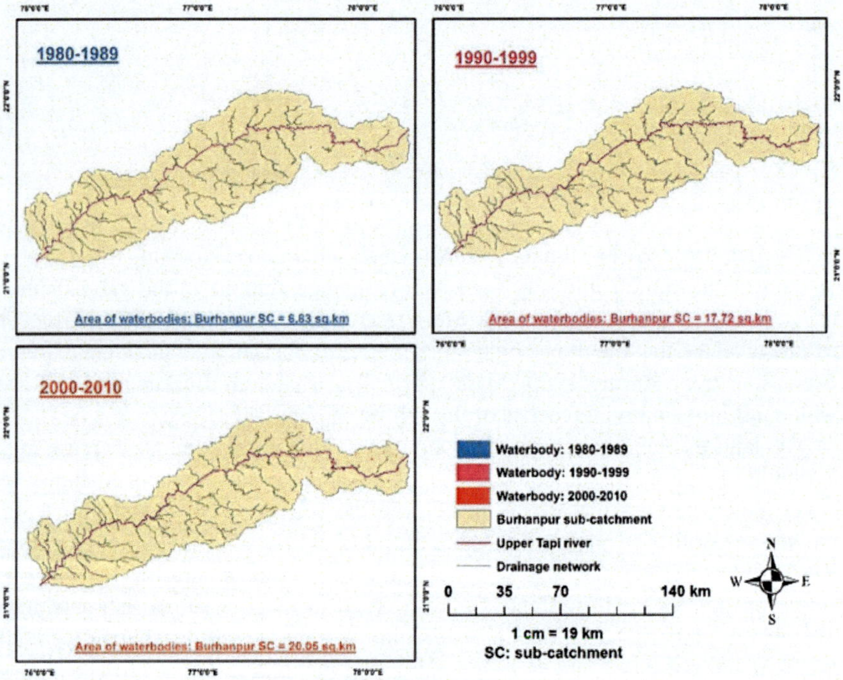

Fig. 13.6 Decadal variation in spatial distribution of waterbodies over the decades in Burhanpur sub-catchment

which has further culminated in the reduction of sediment yield at the outlet of sub-catchment. Also, the reduction in fallow land and scrub lands, which are major sources of unconsolidated sediments, has reduced over the decades and responsible for reduction in the sediment yield.

13.5 Conclusions

The key findings from present study are summarized below:

(a) More than 99% of the sediment load is transported during monsoon months (June–October), wherein, about 92% of the time, the concentration of suspended sediments is below 4 g/L at Burhanpur stream gauging station.
(b) Concentration of the suspended sediment is maximum during the early monsoon season, i.e. end of the June month and throughout the July month.
(c) The mean suspended sediment concentration values were computed as 0.32 g/L during 1980–1989, 0.33 g/L during 1990–1999 and 0.24 g/L during 2000–2010. However, due to the occurrence of high magnitude flows, the average annual sediment yield was significantly higher during the period 1990–1999.
(d) The results of trend analysis of monsoonal rainfall indicated that Burhanpur sub-catchment exhibited decreasing trend during the period 1980–2010. The decreasing trend of rainfall with added influence of land-use land-cover changes within the sub-catchment resulted in lesser sediment yield at the catchment outlet in the recent decade.
(e) The land-use land-cover statistics showed that there has been an increase in the agricultural area in the Burhanpur sub-catchment at the expense of its forest cover, fallow and scrub land. The temporal–spatial distribution of waterbodies, in the form of minor storage structures, were found to be increasing during the period 1980–2010. The reduction of fallow and scrub land and increase in the areas of waterbodies within the sub-catchment were responsible for reduction in sediment yield at the outlet of the catchment during the decade 2000–2010.

Acknowledgements The authors are thankful to the Central Water Commission (CWC), Ministry of Jal Shakti, Government of India for providing necessary financial support through project titled "Morphological Study of Tapi River using Remote Sensing Technique" and data required for the study. The authors are also thankful to Centre of Excellence (CoE) on 'Water Resources and Flood Management', SVNIT-Surat under TEQIP-II for the resourceful help and support.

References

Burylo M, Rey F, Delcros P (2007) Abiotic and biotic factors influencing the early stages of vegetation colonization in restored marly gullies (Southern Alps, France). Ecol Eng 30(3):231–239

Fryirs K (2013) (Dis) Connectivity in catchment sediment cascades: a fresh look at the sediment delivery problem. Earth Surf Proc Land 38(1):30–46

Glymph LM (1954) Studies of sediment yields from watersheds. Int Union Geod Geophys Int Assoc Hydrol 10th Gen Assem Rome, Italy, Part 1:178–191

Hire SP (1999) Geomorphic and hydrologic studies of floods in the Tapi basin. Doctoral dissertation, University of Pune, Pune, India

Hudek C, Burylo M, Rey F (2010) Root system traits of Mahonia aquifolium and its potential use in soil reinforcement in mountain horticultural practices. Sci Hortic 125(3):504–511

Jain MK, Kothyari UC (2000) Estimation of soil erosion and sediment yield using GIS. Hydrol Sci J 45(5):771–786

James LA, Lecce SA (2013) Impacts of land-use and land-cover change on river systems. Treatise Geomorphol 9:768–793

Lü Y, Sun R, Fu B, Wang Y (2012) Carbon retention by check dams: regional scale estimation. Ecol Eng 44:139–146

Mingguo Z, Qiangguo C, Qinjuan C (2008) Modelling the runoff-sediment yield relationship using a proportional function in hilly areas of the Loess Plateau, North China. Geomorphology 93(3):288–301

Ouyang W, Skidmore AK, Hao F, Wang T (2010) Soil erosion dynamics response to landscape pattern. Sci Total Environ 408(6):1358–1366

Parsons AJ, Wainwright J, Brazier RE, Powell DM (2006) Is sediment delivery a fallacy? Earth Surf Proc Land 31(10):1325–1328

Preti F, Forzieri G, Chirico GB (2011) Forest cover influence on regional flood frequency assessment in Mediterranean catchments. Hydrol Earth Syst Sci 15(10):3077

Rey F, Burylo M (2014) Can bioengineering structures made of willow cuttings trap sediment in eroded marly gullies in a Mediterranean mountainous climate? Geomorphology 204:564–572

Shi ZH, Ai L, Li X, Huang XD, Wu GL, Liao W (2013) Partial least-squares regression for linking land-cover patterns to soil erosion and sediment yield in watersheds. J Hydrol 498:165–176

Toy TJ, Foster GR, Renard KG (2002) Soil erosion: processes, prediction, measurement, and control. Wiley, New York, pp 25–43

Trimble SW (1975) A volumetric estimate of man-induced soil erosion on the southern Piedmont Plateau. Prospective Technol Predicting Sediment Yields sources 142–152

Vanmaercke M, Poesen J, Broeckx J, Nyssen J (2014) Sediment yield in Africa. Earth Sci Rev 136:350–368

Walling DE, Fang D (2003) Recent trends in the suspended sediment loads of the world's rivers. Global Planet Change 39(1):111–126

Yan B, Fang NF, Zhang PC, Shi ZH (2013) Impacts of land use change on watershed streamflow and sediment yield: an assessment using hydrologic modelling and partial least squares regression. J Hydrol 484:26–37

Chapter 14
Assessment of Probable Maximum Flood (PMF) Using Hydrologic Model for Probable Maximum Precipitation in Maithon Watershed

Bhanu Sharma and Kalyan Kumar Bhar

Abstract Hydrological modeling is a widely used technique to simulate the hydrological response in a basin due to precipitation for the purpose of basin water management by water resource planners. In hydrological studies, estimating surface runoff in a watershed based on the rate of precipitation received and quantifying discharge at the outlet is important. An analysis has been done for developing hydrologic model for the calculation of probable maximum flood (PMF) using probable maximum precipitation (PMP), over a watershed. PMP is defined as "the greatest depth of precipitation for a given duration that is physically possible over a given storm area at a particular geographical location at a certain time of the year." The focus of the study is firmly on PMP estimates derived through statistical approach (Hershfield's method). The probable maximum flood (PMF) is the flood that can be predicted to be measured using a hydrologic HEC-HMS model from the most extreme combination of critical meteorological and hydrologic conditions that are reasonably possible in particular drainage area. To compute loss, rainfall excess conversion to runoff, and flow routing, methods like initial and constant rate, Snyder's unit hydrograph, and Muskingum routing were chosen, respectively. The study presents a methodology for portraying the PMP estimation for Maithon watershed by analyzing individual steps within the PMP derivation procedure. Calculated PMP value has been used in HEC-HMS model to compute the PMF for Maithon catchment.

Keywords Probable maximum precipitation · Probable maximum flood · Hydrologic model · Hydraulic model · Maithon

B. Sharma (✉)
Department of Hydrology, Indian Institute of Technology, Roorkee 247667, India

K. K. Bhar
Department of Civil Engineering, IIEST, Shibpur 711103, India
e-mail: kalyan@civil.iiests.ac.in

R. Jha et al. (eds.), *Water Resources Management and Reservoir Operation*, Water Science and Technology Library 107,
https://doi.org/10.1007/978-3-030-79400-2_14

14.1 Introduction

Probable maximum precipitation (PMP) is referred as "the greatest depth of the precipitation for a given duration meteorologically possible for a given size storm area at a particular location at a particular time of year, with no allowance made for long-term climatic trends" (WMO 1986). This term is used by many hydrologists all over the world to assess the probable maximum flood (PMF) through which hydraulic structures can be designed. The predominant factor in designing of spillway of a dam is PMP. Machhu II dam located in Gujrat, India, failed in 1979 due to the insufficient spillway capacity and killed 1419 people. There are so many methods related to statistical, dynamic, and empirical approach to find PMP. The magnitude of spatial and temporal distribution has considered for a catchment to find the probable maximum flood at a dam. The designing of dam on the basis of probable maximum flood using probable maximum precipitation is widely used in the world. For different regions of USA, National Weather Service (NWS) published hydrometeorological reports to find PMP. The PMP has various methods to calculate based on the location of watershed, availability of data, and other factors. The long-term data for rain gauge stations has been used in statistical method. This method gives the point value of PMP and needs area reduction curve to vary the area spatially over the area. This paper has used Hershfield's methods to assess PMP, which is a statistical method based on the previous rainfall data.

This PMP value for different location in the watershed is used to assess the PMF for that particular watershed. The peak discharge is used for the design of a hydraulic structure and river headwork after considering hydrological factors. Spillway design flood is used to design the spillway of storage structure which is generally known as the maximum discharge crossing over the spillway without any damages and warning to the stability of structure. The flood that is the result of the most severe combinations including rare combinations of meteorological and hydrological factors is known as probable maximum flood. In hydrologic studies, estimation of surface runoff in a watershed on the basis of precipitation at outlet is the prominent thing. Runoff assessment is a very tedious task due to the variations in parameters. In this study, HEC-HMS model has been used to simulate the rainfall–runoff for a watershed. Simulation of rainfall–runoff model will estimate the runoff. The inappropriate assessment of runoff leads to some problems in adequate management of water resources and reservoir. The peak of the runoff can be assessed based on the rational method, empirical method, and flood frequency method. The watershed models and unit hydrograph techniques calculate flood hydrograph as well as peak flow of the runoff.

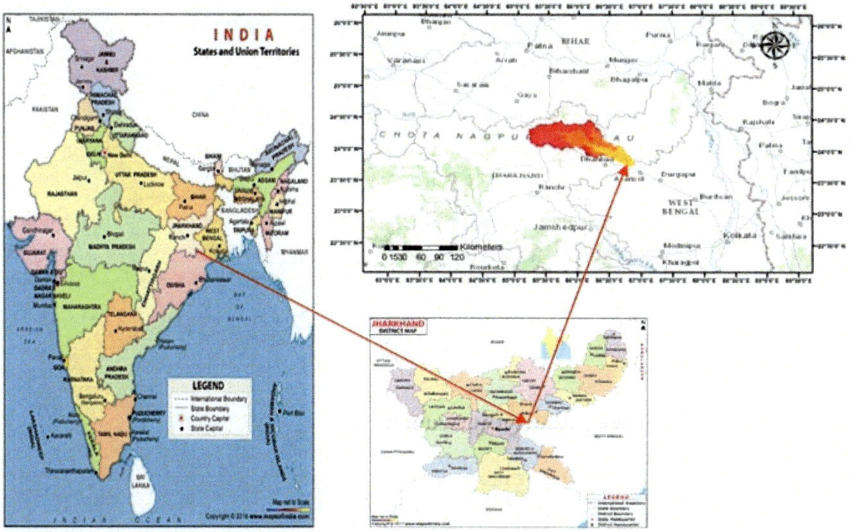

Fig. 14.1 Study area

14.2 Study Area

14.2.1 Geographical Description of Study Area

The present study deals with the hydrologic aspects of an existing dam in India, namely the Maithon dam, which is one of the four dams of the Damodar valley (DV) system. Maithon dam is constructed across river Barakar and the location of dam is at Latitude 23°47′13.06″ north and longitude 86°49′01.44″ east. Barakar river is a tributary of Damodar river. The catchment area upstream of the dam has been estimated as 6391.7 km^2, which spreads over the districts of Dhanbad, Jamtara, Giridih, Kodarma, Hazaribag, and Chatra of the state of Jharkhand. It extends between latitude 23°46′34.12″ north and longitude 85°09′16.26″ east, to latitude 24°32′09.80″ north and longitude 86°53′19.20″ east (Fig. 14.1).

14.2.2 Physiographic Description of Maithon Dam

Maithon dam had completed in 1957 under the supervision of Mr. W.L.Voord, a civil engineer of Tennessee Valley Authority (TVA). Barakar is a seasonal river, 82% of the reservoir filled up by the rainfall in July–September and stored water is used for irrigation by Burdwan area (Table 14.1).

Table 14.1 Physiographic description of Maithon dam

Description	Value
River	Barakar, major tributary of damodar river
Height	56.08 m
Length	4064.35 m—Earthen embankment 362.41 m concrete overflow section
Gross storage capacity	1093.54 mm^3
Live storage capacity	441.64 mm^3
Full reservoir level	152.40 m
Minimum draw down level	132.59 m
Dead storage level	132.59 m
Dam top level	156.06 m
Crest level	140.21 m
Spillway type	Ogee
Type of gates	Radial
Size	12.5 m (h) × 12.19 m (w)
No. of bays	12 nos

14.3 Materials and Methodology

In this present study, the rainfall data (1994–2015) collected from Indian Meteorological Department (IMD) was used to prepare the rainfall distribution map. Digital elevation model (DEM) and the satellite images are used as the primary source of data. The digital elevation model (DEM) has been taken from Advanced Space borne Thermal Emission and Reflection Radiometer (ASTER) that is a free product of United States Geological Survey. It has a horizontal spatial resolution of 30 m and a vertical accuracy of more than 10 m. The study area comprises of Barakar river catchment, which was delineated by automatic catchment delineation from DEM in ArcGIS 10. Different rainfall distribution map and PMP distribution maps are created using ArcGIS software.

14.3.1 Delineation of Catchment

Catchment is an area or region bounded peripherally by a divide and eventually drained to a single water source or body of water. A series of operations has been carried out to extract the drainage network and the catchment with the help of the

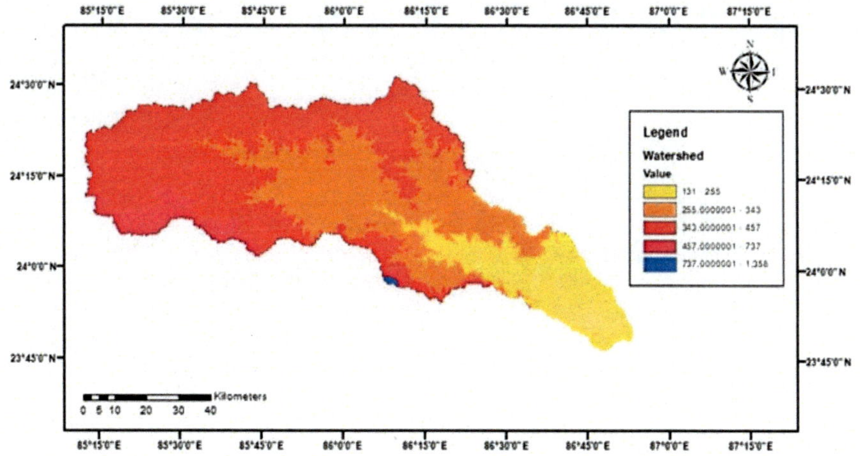

Fig. 14.2 Maithon watershed

software ArcGIS 10.0. There are so many steps like fill, flow direction, flow accumulation, stream network, stream feature, and snap pour point had been performed (Fig. 14.2).

14.3.2 Preparation of Sub-catchment

Each of the tributaries of the main river or lake system has its own drainage area, known as a sub-catchment. To delineate the sub-catchment, previous procedure is used with inclusion of multiple pour point. Figure shows the delineated sub-catchment. The following table shows the area of different sub-catchments (Fig. 14.3; Table 14.2).

14.3.3 Rainfall Distribution Maps

The first step is to identify the rain gauge stations covering the entire catchment. The namc of rain gauge stations is Barkisuriya, Maithon, Nandadih, Palganj, Giridh, Jamtara, Padma, Jamua, Barhi, Sillai Chalk, Tilaiya, and Dumri under the Maithon catchment. On the basis of rainfall data, the rainfall map has been made, which shows the average value of rainfall occurring in the years from 1992 to 2017. After identifying the rain gauge station, the spatial variation of rainfall has been shown with the help of rainfall distribution map over the catchment area made with the help of inverse distance weighting (IDW) interpolation technique (Fig. 14.4).

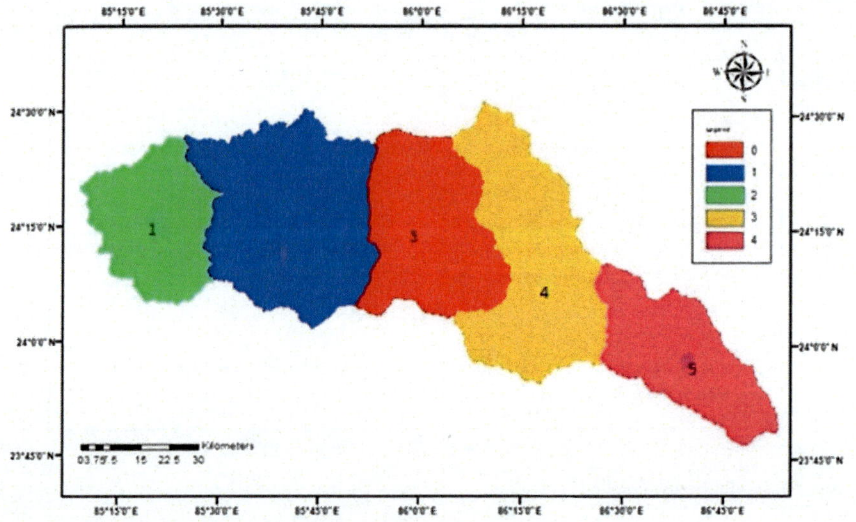

Fig. 14.3 Sub-catchments of Maithon watershed

Table 14.2 Area of sub-catchments

Sub-catchment	Area (Sq. Km.)
1	1031.80
2	1451.95
3	1225.45
4	1768.83
5	910.66
Total area	6388.69

14.4 Probable Maximum Precipitation

PMP is defined as "the greatest depth of precipitation for a given duration that is physically possible over a given storm area at a particular geographical location at a certain time of the year." The statistical method, often called Hershfield method (1961), is more commonly used and can be applied, if long-term precipitation data is available.

The Hershfield method is based on average precipitation and standard precipitation deviation, similar to the frequency factor method of Chow (1951), which is expressed as (Fig. 14.5; Table 14.3):

$$P = \bar{X} + K_m S_n \tag{14.1}$$

where

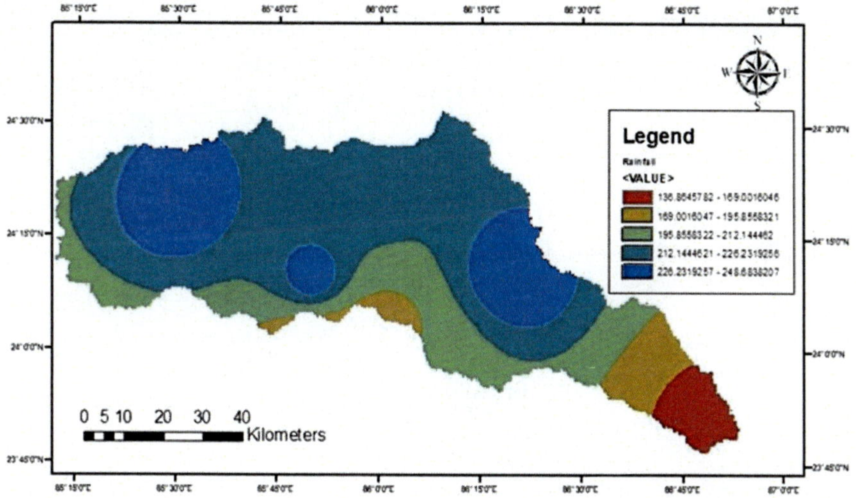

Fig. 14.4 Rainfall map

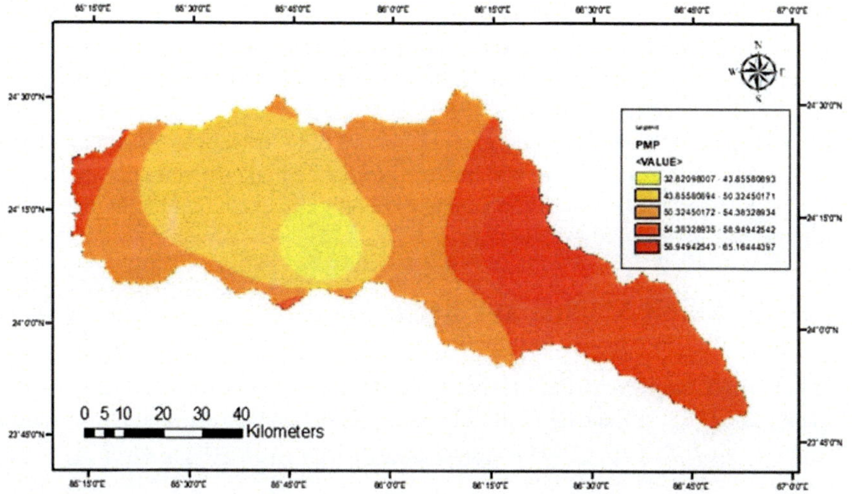

Fig. 14.5 PMP maps

$$\bar{X} = \frac{1}{n}\sum_{i=1}^{n} X_i \tag{14.2}$$

$$S_n = \sqrt{\frac{\sum_{i=1}^{n}\left(X_I - \bar{X}\right)^2}{n-1}} \tag{14.3}$$

Table 14.3 PMP values

Station	PMP values (cm)
Barkisuriya	32.81
Maithon	46.78
Nandadih	47.96
Palganj	51.08
Giridh	65.16
Jamtara	58.09
Padma	65.14
Jamua	53.10
Barhi	59.10
Sillai Chalk	64.09
Tilaiya	47.06
Dumri	54.04

$$K_m = \frac{X_i - X_{n-1}}{\sigma_{n-1}} \quad (14.4)$$

$\bar{X}$ Sample mean of n annual maximum precipitation.
S_n Sample standard deviation.
K_m Frequency factor.
X_i Highest observed annual maximum rainfall in the series.
X_{n-1} Mean of the annual maximum, excluding the highest value.
σ_{n-1} Standard deviation of the annual precipitation, excluding the highest value.

14.5 Probable Maximum Flood (PMF)

Probable maximum flood (PMF) is calculated by HEC-HMS hydrologic model which is being used. To compute loss, rainfall excess conversion to runoff and flow routing, methods like initial and constant rate, Snyder's unit hydrograph and Muskingum routing were chosen, respectively.

14.5.1 Physiographic Parameter of the Sub-catchments

The physiographic parameters of the mentioned above were extracted for each of the five sub-catchments using DEM by ArcGIS (Table 14.4).

Table 14.4 Sub-catchment centroid, longest stream

Sub-catchment	Area (km^2)	Length of longest stream (km)	Length of stream to centroid (km)
R1	910.66	65.97	32.80
R2	1768.83	92.34	39.96
R3	1225.45	75.56	30.36
R4	1451.95	93.19	38.47
R5	1031.80	79.61	29.42
$\sum$	6388.69		

Table 14.5 Basin lag time

Sub-catchment	Length of longest stream (km)	Length of stream to centroid (km)	Basin lag time (h)
R1	65.97	32.80	55.071
R2	92.34	39.96	64.634
R3	75.56	30.36	56.045
R4	93.19	38.47	64.078
R5	79.61	29.42	56.39

14.5.2 Snyder's Method

Snyder (1938) developed a synthetic UH based on a study of watersheds in basins ranging from 10 to 10,000 mi^2 (Table 14.5).

$$t_{\text{lag}} = CC_t(LL_c)^{0.3} \tag{14.5}$$

t_{lag} Basin lag time.
C Conversion constant (2.75).
C_t UH peaking coefficient (1.8–2.2).
L Length of the main stream from the outlet to the divide (km).
L_c Length along main stream to appoint nearest the watershed centroid (km).

14.5.3 Design Loss Rate

For assessment of probable maximum floods, a loss rate of 0.1 cm/hour is currently used, as the influence of soil, land use, etc., have the minimal influences under such extreme conditions. So, a loss rate of 0.1 cm/hour was adopted for the current study.

14.5.4 Base Flow

For the purpose of this study, the base flow rate has been adopted as 0.045 $m^3/s/km^2$.

14.5.5 Muskingum Parameters

See Table 14.6.

14.5.6 HEC-HMS Model

See Fig. 14.6.

Table 14.6 Muskingum coefficient

Reach	Proportionality coefficient (K)	Weighing factor (X)
R1	8.2	0.2
R2	5.9	0.2
R3	3.4	0.2
R4	5.8	0.2

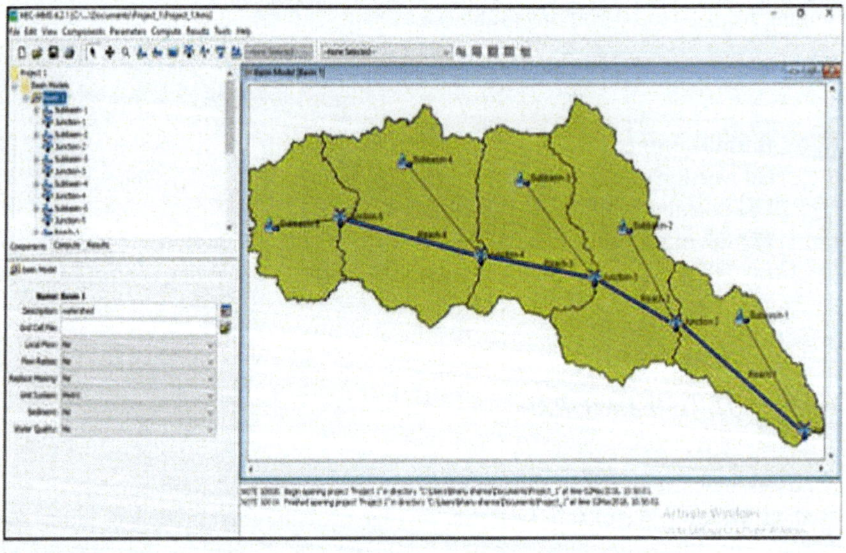

Fig. 14.6 Basin model

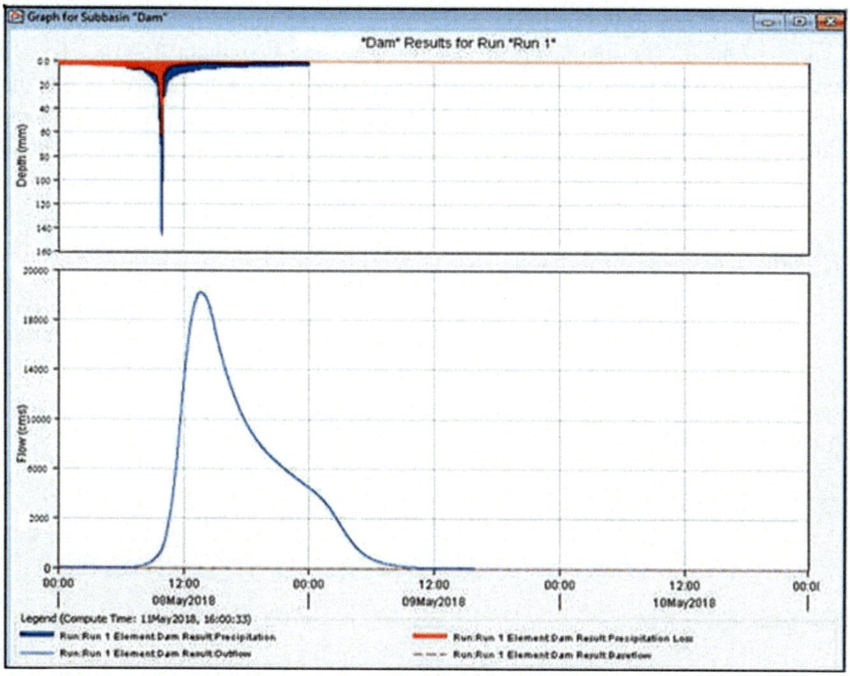

Fig. 14.7 Final outflow hydrograph

14.5.7 Output

See Fig. 14.7.

14.6 Conclusion

The one-day probable maximum precipitation for Maithon catchment is 43.02 cm using Hershfield's method for 12 rain gauge stations. We can also see the point PMP values for different stations given in the table above. Maithon, Nandadih, and Tilaiya are having the value in the range of 40–50 cm. Palganj, Jamtara, Jamua, Barhi, and Dumri are having the value in the range of 50–60 cm. Giridh, Padma, and Sillai chalk having the values in the range of 60–70 cm. statistical method has the advantage of taking into account actual precipitation data and its application is simple and fast. A long record will yield generally more reliable PMP estimates than will a short record of comparable quality. The peak outflow at the outlet of Maithon catchment is 18751 m^3/s. The spillways are originally designed to sustain the discharge of 13,592 m^3/s. Here, we can clearly see that the design flood has increased by 37% for Maithon watershed. The variation in rainfall due to climate change and due to the silt charge accumulated in the reservoir is the prominent reason for increment in design flood. The increment in value of probable maximum flood evolves the problem like dam break which leads to the flooding in downstream area of dam that causes financial loss, loss of life, and decrement in soil fertility. This study can have the prominent role in dam break scenario and evacuation policies during dam break.

References

Chow VT (1951) A general formula for hydrologic frequency analysis. Trans Am Geophys Union 32(2):231–237
Snyder FF (1938) Synthetic unit-graphs. Eos. Trans Am Geophys Union 19(1):447–454
WMO (1986) Manual on estimation of probable maximum precipitation (PMP). ISBN 978-92-63-11045-9

Chapter 15
Simulating Failure of Indravati Dam Using Mike 11 and the Propagation of Breached Outflow

Aditya Harikumar, Sachin Dhiman, and K. C. Patra

Abstract Dams are invariably used for multipurpose benefits to the society such as generation of hydropower, irrigation, water supply, flood control, and others, but with associated high risk. As it retains huge amount of water, there is always a risk of formation of breach. There is a need to study the breaching aspects of a dam. In this paper, a hydraulic model named MIKE11 developed by Danish Hydraulic Institute (DHI) is used to carry out simulation of flood resulting from the failure of Indravati multipurpose dam in the Odisha state. Propagation of the flood wave at its downstream reaches is studied, and the consequent inundation map of the downstream submerged areas is prepared. Digital elevation map of the study area is obtained from USGS site and is converted into ARCII with the help of ArcGIS software. Simulations carried out posed a challenge due to the availability of low-resolution topographic data, stiff slope of the channel, obstructions at the downstream side of dam, and other factors.

Keywords ArcGIS · Indravati dam · Dam breach · Flood inundation · MIKE11

15.1 Introduction

Dams are the basic piece of frameworks to the general public that add to social advancement and success. They serve too many beneficial purposes that are essential for the growth of the society; at the same time, dams hold a potential danger of disappointment because of numerous specialized security issues and threats. Disappointments are viewed as one of the real "low-likelihood" occasions. The surges coming due to dam failures can prompt catastrophes with gigantic death toll and property, particularly in thickly populated territories.

Past calamities resulting from dam disappointments are specifically identified with the clearing time when a dam failure happens (Wahl 1998). It is along these lines critical to comprehend the procedure of dam rupturing, and if conceivable, to consider key breaking parameters required to display the dam rupturing process. Be

A. Harikumar (✉) · S. Dhiman · K. C. Patra
Department of Civil Engineering, NIT, Rourkela, India

R. Jha et al. (eds.), *Water Resources Management and Reservoir Operation*,
Water Science and Technology Library 107,
https://doi.org/10.1007/978-3-030-79400-2_15

that as it may, many existing dams still posture expanding risks to the downstream territories because of basic crumbling, lacking outline, defective development, and poor activity and upkeep. These dams are alluded to as upset dams.

Hanson et al. (2005) clarified that the breach development rate has a critical effect on the pinnacle release from a dam failure. In order to understand the breach development rate, one has to keep a check on the breach initiation and formation times.

Wahl (1998) had given a clear cut explanation about two types of times associated with dam breach. The breach initiation time, which can be utilized to start an early cautioning, is the span from the beginning of overtopping stream over the dam dissolving downstream slope to the beginning of bringing down of the peak of the dam interfacing stored water to the broke outpouring. Breach development time is the span beginning from the bringing down of the bank peak and consummation with the total draw off the reservoir water.

Over the last two decades, some software have been developed for modeling of dam break that includes HEC-RAS, MIKE11, MIKE 21, and MIKE FLOOD. Indubitably, quality, and dependability of the counterfeit processes for every component of the river flood boost the quality and authenticity of the created river flood danger maps.

For successful running of any software model, it requires a correct representation of channel of the downstream corresponding floodplains and other data as accurately as possible to predict the flood magnitude and level of water along the channel after the dam break (Gichamo et al. 2012).

For this purpose, DEM file is required for the study area. Apart from the topographical data, satellite images also provide the exact information along with the different structures present in the area of interest. This is very much useful to get an idea of the area that can get inundated after the burst of a particular dam on the upstream (Qi and Altinakar 2011).

Many researchers have worked on this field to get an idea of the peak outflow from the dams while connecting it to the different parameters [(SCS 1981), Singh and Snorrason (1984), Froehlich (1995 and 2008)]. Froehlich (1995) had collected data from 63 dam failure conditions to get an idea of the average width of breach. In order to get an idea about the formation time of the breach, he used data from 21 real cases of dam failure. Majority of the works are focussed to two parameters, i.e., average width of breach and formation time of the breach, as these are more sensitive to failure peak during breaching of a dam.

In this paper, a numerical simulation of the failure of Indravati dam is carried out by employing Danish Hydraulic Institute's (DHI)—MIKE 11. In this study, the effect of overtopping failure is given attention through Saint–Venant equations. A reliable and accurate mathematical model needs to be developed for breach evolution for analyzing the impact of dam-break floods on downstream regions (Dhiman and Patra 2018).

15.2 Study Area

15.2.1 Salient Features of Indravati Dam

River Indravati originates in the eastern mountainous regions of India in the state of Odisha. It is an east flowing river that enters into Chhattisgarh state, finally joining the Godavari river. Indravati river basin lies between latitudes 18°43′25″ N to 19°26′46″ N and longitudes 80°16′19″ E to 83°07′10″ E. Major tributaries of this river are Bhaskal, Narangi, Nibra, Kotri, Bandia, and Nandiraj rivers. Figure 15.1 shows the Indravati basin and dam while the other feature of Indravati dam is given in Table 15.1. Figure 15.2 depicts a schematic diagram of the Indravati river system.

Fig. 15.1 Map showing Indravati river basin and the dam

Table 15.1 Salient features of Indravati dam

Dam location and properties		Reservoir capacity	
River	Indravati	Gross storage capacity at F.R.L	2307 M cumec
State	Odisha	Dead storage capacity	851.94 M cumec
Latitude & longitude	19°16′34.8″ N and 82°49′42.4″ E	Live storage capacity	1455.76 M cumec
Catchment area	2630 sq. km	Full reservoir level	R.L. 642 m
Length	539 m	Maximum water level	R.L. 643 m
Height	45 m	Min draw down level	R.L. 625 m
Dam top crest level	R.L. 645 m	Maximum flood discharge	23,030 cumec
Deepest bed level	R.L. 600 m		

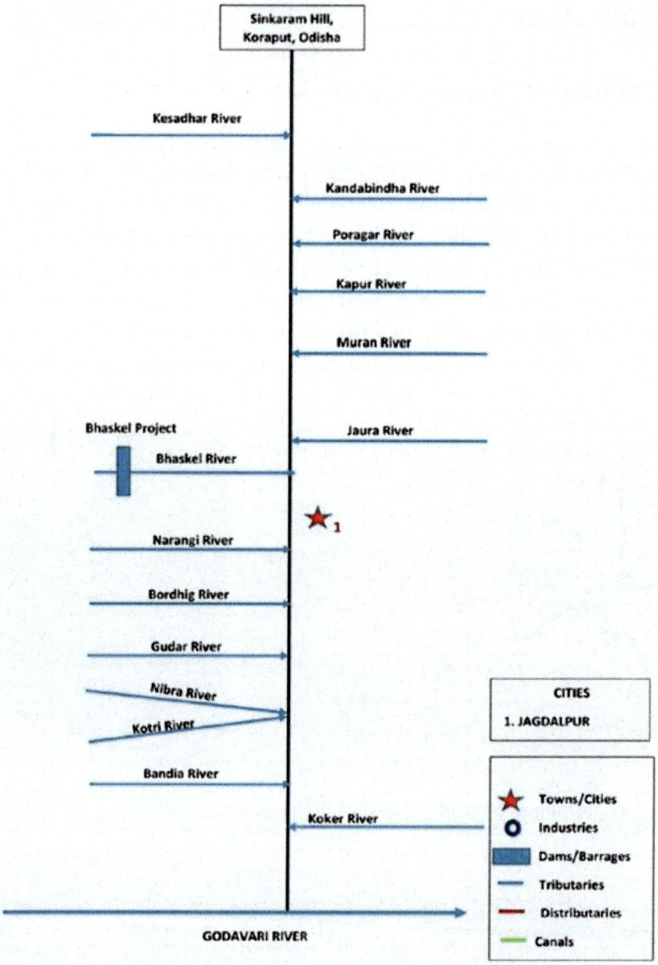

Fig. 15.2 Schematic diagram of Indravati river basin

15.3 Dam Breach Parameters

It is one of the most important inputs which have to be carefully considered in a dam-break analysis. Usually earthen dams take more time for its complete failure as compared to a masonry dam. The failure time of earthen dam is 0.1 to 1 h and that of masonry dam is 0.1 to 0.2 h (Fread 2006).Table 15.2 shows the guidelines followed during dam-break analysis as recommended by the UK and the federal energy regulatory commission (FERC) of the USA.

There are various ways in which breaches can be formed on embankment dams, and it is pretty difficult to find out the extent of the erosion with the help of stringent mathematical formulas. Breach formation in embankment dams due to overtopping

Table 15.2 Guidelines followed during dam break by UK and federal energy regulatory commission (FERC) of USA

Dam type	Average breach width	Failure time hrs	Breach side slope H:1 V	Agency
Eaithen/Rockfill	(0.5–5.0) × HD (1.0–5.0) × HD (2.0–5.0) × HD	0.5–4.0 0.1–1.0 0.1–1.0	0–1.0 0–1.0 0–1.0	USACE (2007) FERC (1988) NWS(Fread 2006)
Concrete gravity	Multiple Monoliths Usually ≤ 0.5 L Usually ≤ 0.5 L	0.1–0.5 0.1–0.3 0.1–0.2	Vertical Vertical Vertical	USACE (2007) FERC NWS (Fread 2006)

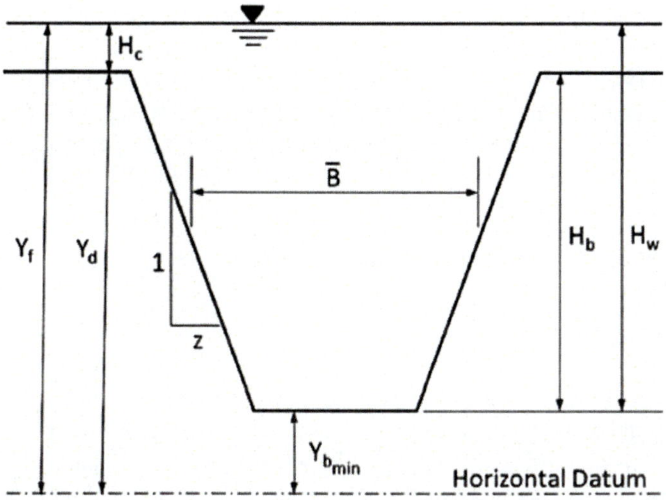

Fig. 15.3 Dimensions of a trapezoidal dam breach approximation, (height H_b, width $\bar{B}$ and ratio of side slope z (H:V), water in the reservoir Y_f)

floodwaters has been counterfeit using complex 2D depth averaged flow models connected with slope failure method and soil erosion (Froehlich 2004; Faeh 2007). Usually in practice, breach is presumed to take the shape of a trapezium as shown in Fig. 15.3 (Fread 1984; USACE 1978).

15.4 Methodology

The DEM file of the study area is converted into arcii format to import it into the mike hydro setup. The first step in mike hydro is to create a simulation file to save the progress of the work. In most of the dam-break cases, there is a single or multiple channels, a reservoir, the dam and its structures like spillways and others. At first,

Table 15.3 Dam breach parameters

Dam crest elevation	645 m	Breach width at end	250 m
Reservoir level at breach	644.3 m	Breach slope	0:1 (vertical)
Breach width at start	10 m		

the network file is created, then the branches of the reservoir and downstream river are created. While creating/giving the cross section, it is mandatory to give the first chainage section of the reservoir to mark it as the storage. After that the boundary file is created, inflow at the upstream is required for the simulation. Time series for water level and discharge is to be made, after which HD parameters are to be created. The HD module makes use of an implicit method for the calculation of unsteady flow in rivers. After all this procedure, the setup is run and for this a simulation editor is created. While considering the dam breach modeling, one of the greatest uncertainties is the simulation of the breach (Wurbs 1987). Depending upon the size of the reservoir, the parameters importance varies. The dam breach parameters used are shown in Table 15.3.

15.5 Model Setup

The first step is the creation of Indravati river in the network editor. The river extends up to a stretch of 138.9 km at its downstream. Up to a length of 50,000 m, the river cross sections are provided at spacing of 500 m and after that and till 138,900 m it is provided at a spacing of 1000 m. The dam-break arrangement is defined at 130 m chainage from the starting point. At chainage "0", the reservoir is modeled. The PMF is considered as the inflow to the reservoir. At chainage point of 138,900 m, boundary conditions are applied. The q/h values are generated automatically from manning's formula. Table 15.4 gives the *q/h* values, and Fig. 15.4 shows the rating curve obtained from MIKE 11. Manning's roughness coefficient for the whole river is taken as 0.022 following Chow and V.T. (1959) guidelines.

Table 15.4 Stage discharge values

Stage (h) in m	Discharge (q) in m^3/s	Stage (h) in m	Discharge (q) in m^3/s
537.6	0	541.7	6069
538	22	542	7636
539	335.6	545.7	29,791.3
540	1879.4	550	61,702.7
541	3765.7		

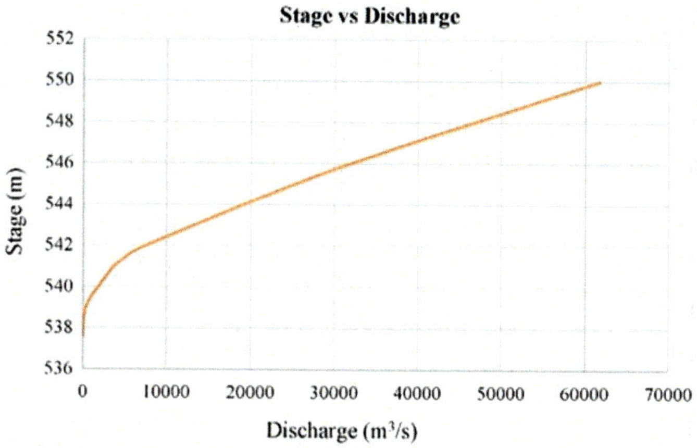

Fig. 15.4 Rating curve

15.6 Results and Analysis

During dam-break analysis, the most critical state is when the water in the reservoir is at full reservoir level (FRL) and a PMF infringes the reservoir. The value of PMF considered is 23,030 cumec. Dam breach is initiated at 25.3 h from the beginning of PMF, and the maximum water level attained at that time is 645.201 m. The maximum value of discharge flowing out of the dam is 33,218.4 cumecs, which is 1.44 times more than the PMF. This maximum discharge is attained at 6 h from the beginning of the dam break. Velocity of water coming out at this point is 10.6 m/s. The corresponding value of breach parameters is given in Table 15.5, while Table 15.6 shows the statistics of dam breach. Figures 15.5 and 15.6 show the discharge versus the time graph and reservoir water level vs the time graph, respectively.

15.6.1 Flood Routing

For the purpose of flood routing at its downstream reaches, 8 chainage points are considered at locations of 1.5, 5, 7.5, 10.5, 25, 55, 100, and 120 km d/s of the reservoir. At the dam, the value of peak discharge coming out after breaching process commences is 33,218.4 cumec. This discharge comes out from the dam after a time

Table 15.5 Breach parameters at maximum discharge

Level of breach (m)	620
Depth in breach (m)	11.97
Breach bottom width (m)	250
Breach width at crest (m)	300

Table 15.6 Statistics of dam breach

Time (h)	Q in breach (m^3/s)	V in breach (m/s)	Reservoir water level (m)	Level of breach (m)	Depth in breach (m)	Breach bottom width (m)	Breach width at crest (m)
25.3	12.2	2.215	645.201	644.3	0.525	10	11.4
25.5	38.4	2.806	645.205	643.7	0.844	15.3	17.8
26	332.1	4.355	645.217	641.7	2.038	35.3	41.9
26.5	1034.2	5.474	645.211	639.6	3.221	55.4	66.0
27	2237.3	6.387	645.174	637.67	4.386	75.4	90.1
27.5	3999.5	7.168	645.09	635.64	5.522	95.5	114.2
28	6346.9	7.849	644.938	633.6	6.618	115.5	138.3
28.5	9272.3	8.447	644.697	631.5	7.661	135.6	162.4
29	12,736.7	8.974	644.346	629.51	8.638	155.6	186.5
29.5	16,669.3	9.434	643.864	627.51	9.538	175.7	210.67
30	20,968.2	9.833	643.229	625.41	10.347	195.7	234.7
30.5	25,503.2	10.17	642.423	623.42	11.054	215.8	258.6
31	30,112.8	10.447	641.422	621.42	11.646	235.8	282.9
31.3	**33,218.4**	**10.597**	**640.551**	**620**	**11.967**	**250**	**300**
31.4	33,008.8	10.576	640.468	620	11.917	250	300.0

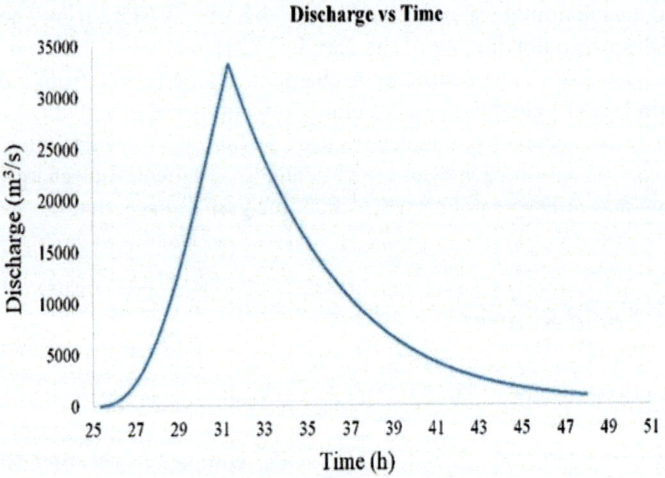

Fig. 15.5 Discharge versus time at the dam site

of 6 h from the start of the breach. At 1.5 km d/s the value of maximum discharge is 33,067.48 cumec. The time of arrival of flood is just 1 min from the beginning of flood. At 5 km d/s, the value of maximum discharge is 32,899.44 cumec, while the arrival time of peak value of flood is 9 min. At 7.5 km d/s, the value of maximum

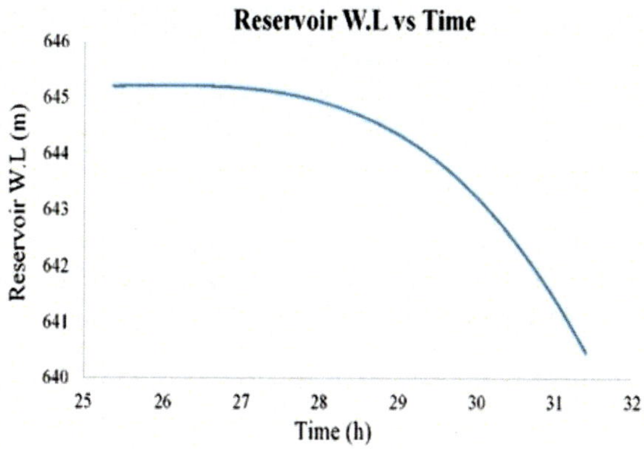

Fig. 15.6 Reservoir w.l versus time

discharge is 32,704.8 cumecs, while the arrival time of peak value of flood is 12 min. At 10.5 km d/s, the value of maximum discharge is 32,453.44 cumecs, and its arrival time is 17 min. At 25 km d/s, the value of maximum discharge is 29,239.14 cumecs, while its arrival time is 1 h 3 min. At 55 km d/s, the value of maximum discharge is 23,895.56 cumecs, while its arrival time is 3 h 23 min. At 100 km d/s, the value of maximum discharge is 16,526.59 cumecs, and its arrival time is 8 h 30 min. At 120 km d/s, the value of maximum discharge is 13,693.69 cumecs, and its arrival time is 11 h 3 min.

15.6.2 Longitudinal Profile of the Bed

Figure 15.7 depicts the longitudinal profile of the bed of river Indravati after dam break along with the water level attained at different downstream locations resulting

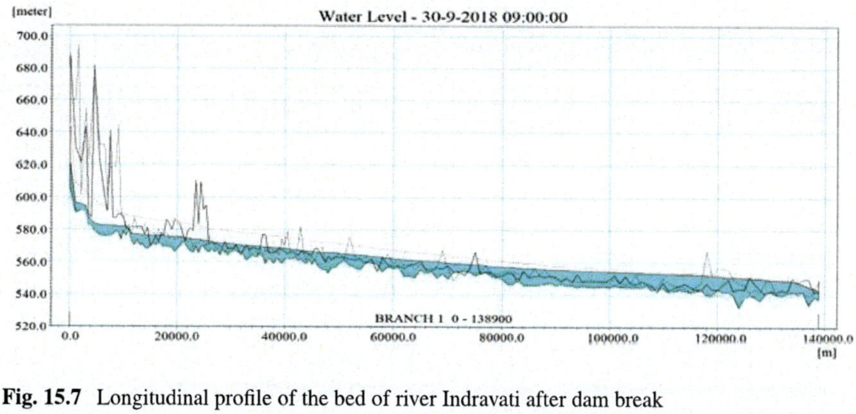

Fig. 15.7 Longitudinal profile of the bed of river Indravati after dam break

from the dam break. Figure 15.8 represents the time series discharge of the river at midpoint of selected sections.

Level of water at all the 8 chainages is shown in Table 15.7. Figure 15.9 depicts the time series water level of all the 8 chainages.

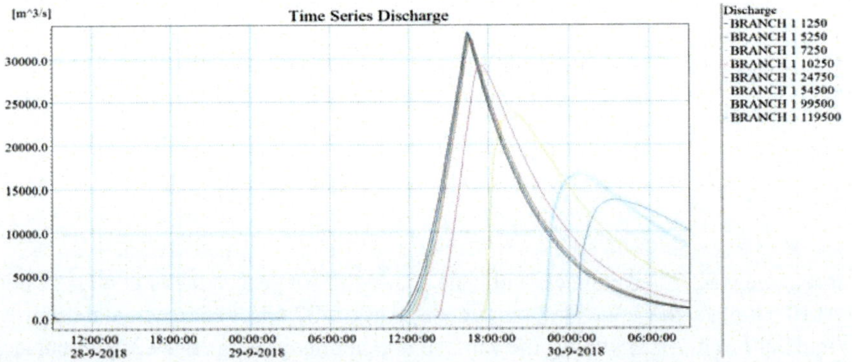

Fig. 15.8 Time series discharge

Table 15.7 Level of water along with the time of occurrence at each chainages

Distance d/s (km)	w.l. (m)	Max. w.l. time	Distance d/s (km)	w.l. (m)	Max. w.l. time
1.5	606.87	29/09/2018 16:24:0	55	569.67	29/09/2018 20:31:5
5	596.69	29/09/2018 16:30:0	100	556.34	30/09/2018 02:24:0
7.5	593.94	29/09/2018 16:33:5	120	552.69	30/09/2018 04:51:5
10.5	589.08	29/09/2018 16:45:5			
25	581.45	29/09/2018 17:51:5			

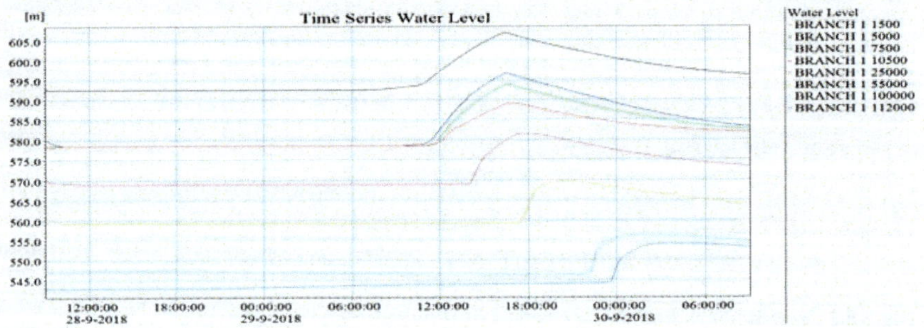

Fig. 15.9 Time series water level

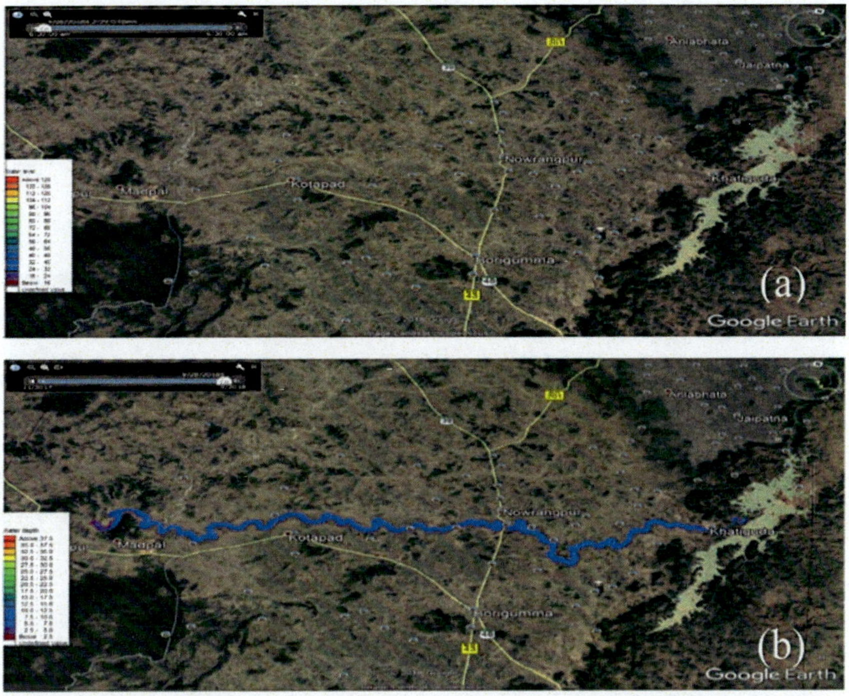

Fig. 15.10 Flood inundation map of Indravati river (a) before and (b) after flooding

15.7 Flood Maps

Based on the contour maps available, the area which comes under the effect of flood can be obtained. This information is shown in Fig. 15.10a and b as before and after images. Figure 15.11a and b shows the before and after image of one of the area along the course of the river where the flood advances into the adjacent flooding plains.

15.8 Emergency Action Plan (EAP)

EAP is a written document given by the dam authorities specifying the plan of action that are required to be taken for the protection of a dam or to control the loss of life and property. It is not like a provisional plan which stresses on the maintenance and corrective construction of the dam rather it helps in the simplification of the problems of dam safety. The central dam safety organization (CDSO) under the central water

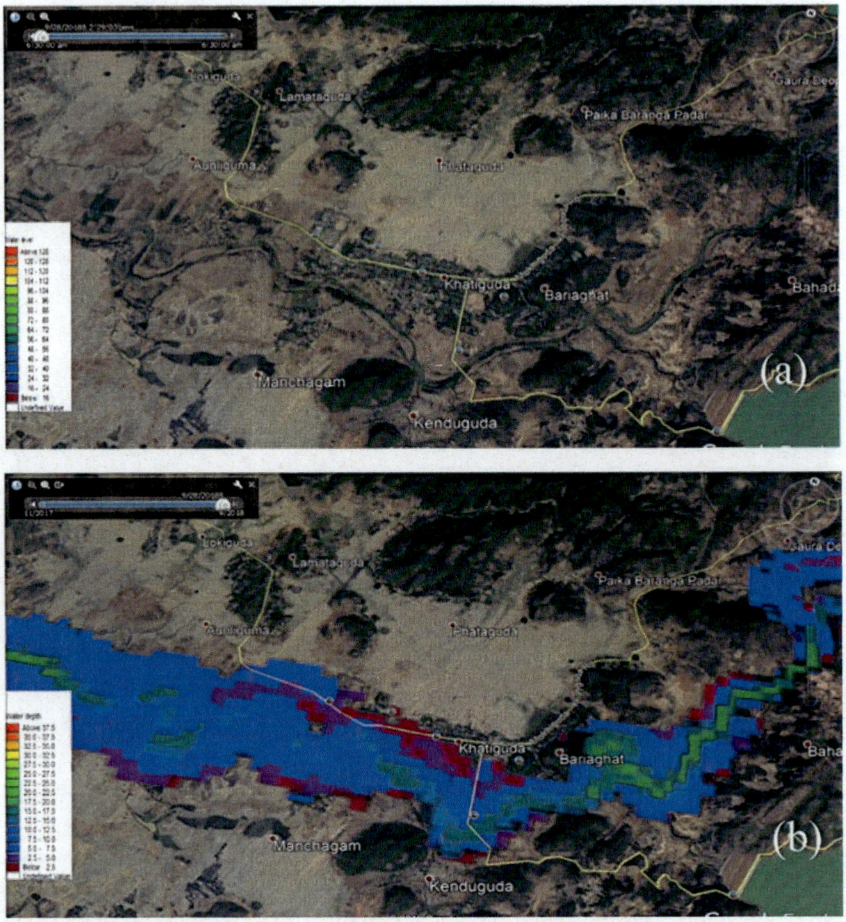

Fig. 15.11 Flood map at Khatiguda which is 5 km d/s of dam (a) before and (b) after flooding

commission (CWC) aims in simplifying the problems in dam safety. In the current situation, a severe weather condition is taken in which heavy rainfall is also taken into consideration. Breaching of the Indravati dam takes place when the water level reaches a height of 645.201 m which is above the top of the dam. Dam breach is initiated at 25.3 h from the beginning of PMF. The maximum value of discharge flowing out of the dam is 33,218.4 cumec which is 1.44 times more than the PMF and it occurs at 31.3 h which signifies the breach in the dam within 6 h. From Table 15.8, we can get to know the discharge at all the 8 chainages and their respective time of arrival. There are three levels of danger signal such as emergency level-1(yellow), emergency level-2 (orange), and emergency level-3 (red). These are described on the basis of breach time available and the locality affected by the flood.

From Table 15.8, it is evident that some parts of the river reach area come under extreme danger zone, i.e., locations varying from 1.25 to 25 km from dam site. These

Table 15.8 Discharge along with the time of occurrence and emergency level at each chainages

Distance d/s (km)	Discharge (m^3/s)	Max. Q. Time	Emergency level
1.25	33,067.48	29/09/2018 16:21:5	Red
5.25	32,899.44	29/09/2018 16:30:0	Red
7.25	32,704.80	29/09/2018 16:33:5	Red
10.25	32,453.44	29/09/2018 16:38:0	Red
24.75	29,239.14	29/09/2018 17:24:0	Red
54.5	23,895.56	29/09/2018 19:44:0	Orange
99.5	16,526.56	30/09/2018 00:51:5	Yellow
119.5	13,693.69	30/09/2018 03:24:0	Yellow

areas are very much affected because during the propagation of the wave there is a fair chance that the community living in these areas would be affected a lot.

15.9 Conclusions

Here, in this paper, a hypothetical failure of Indravati reservoir is simulated using MIKE 11. The dam has a height of 45 m with gross storage capacity of 2300 m cumecs. Effects of dam break at the downstream side of the dam are studied with the help of flood hydrograph, duration of the flood, water level, and velocity of propagation of flood wave. From the modeling, the following conclusions can be drawn.

- Maximum flood discharge from the failure of the Indravati dam is 33,218.4 cumec, which is 1.44 times more than the PMF.
- The dam-break results can be utilized to prepare maps for an emergency action plan that can help communities in arranging future improvements in zones that are flooding prone.
- The time series water level and discharge at different c/s (1.5, 5, 7.5, 10.5, 25, 55, 100, and 120 km) are obtained that suggests the flood-prone areas resulting from the dam breach.

References

Chow VT (1959) Open-channel hydraulics. McGraw-Hill, New York, pp 680

DHI (2008) MIKE 11- A modeling system for rivers and channels, user's manual. Danish Hydraulic Institute, Horsholm, Denmark

Dhiman S, Patra KC (2018) Experimental study of embankment breach based on its soil properties. ISH J Hydraul Eng. https://doi.org/10.1080/09715010.2018.1474500

Faeh R (2007) Numerical modeling of breach erosion of river embankments. J Hydraul Eng 133(9):1000–1009
Federal Power Commission (1984) Hydroelectric power resources of the United States, developed and undeveloped. Federal Power Commission
Fread DL (1984) DAMBRK—the nws dam-break flood forecasting model. National weather service hydrological technical note no. 4, hydrologic research laboratory, silver spring, md
Fread DL (2006) Dam failure analysis. ASDSO Adv Technical Sem
Froehlich DC (1995) Peak outflow from breached embankment dam. J Water Resour Plann Manage 121(1):90–97
Froehlich DC (2004) Two-dimensional model for embankment dam breach formation and flood wave generation. In: Proceeding of the dam safety 2004, 21st annual conference of the association of state dam safety officials, association of state dam safety officials, Lexington, ky
Froehlich DC (2008) Embankment dam breach parameters and their uncertainties. J Hydraul Eng 134(12):1708–1721
Gichamo TZ, Popescu I, Jonoski A, Solomatine D (2012) River cross-section extraction from the aster global dem for flood modelling. Environ Model Softw 31:37–46
Hanson GJ, Cook KR, Hunt SL (2005) Physical modelling of overtopping erosion and breach formation of cohesive embankments. Trans ASAE 48(5):1783–1794
Qi H, Altinakar MS (2011) A GIS-based decision support system for integrated flood management under uncertainty with two dimensional numerical simulations. Environ Model Softw 26(6):817–821
SCS (1981) Simplified dam-breach routing procedure. Technical release no. 66 (rev. 1). USDA-NRCS, Washington, D.C.
Singh KP, Snorrason A (1984) Sensitivity of outflow peaks and flood stages to the selection of dam breach parameters and simulation models. J Hydrol 68(1):295–310
U.S. Army Corps of Engineers (USACE) (1978) Flood hydrograph package HEC-1 user's manual for dam safety investigations. Hydrologic Engineering Center, Davis, Calif
U.S. Army Corps of Engineers (USACE) (2007) Performance evaluation of the New Orleans and southeast Louisiana hurricane protection system: final report of the interagency performance evaluation task force, vol III
Wahl TL (1998) Prediction of embankment dam breach parameters—a literature review and needs assessment. Dam safety rep. No. Dso-98–004, U.S. department of the interior, bureau of reclamation, Denver
Wurbs RA (1987) Dam-breach flood wave models. J Hydraul Eng 113(1):29–46

Chapter 16
Optimization of Water Allocation for Ukai Reservoir Using Elitist TLBO

Vijendra Kumar and S. M. Yadav

Abstract The optimal reservoir policy is a complex problem to optimize. This paper introduces the improved teaching learning-based optimization (TLBO) by introducing the elitist concept, in order to improve the convergence, global searchability and faster evolution process. The objective of the present study is to maximize the water allocation for Ukai reservoir, India, to supply water for irrigation, domestic and industrial uses at different dependable inflow. Elitist teaching learning-based optimization (ETLBO) algorithm has been used to optimize water allocation, using four different models having dependable inflow as 60, 65, 70 and 75%. The results from ETLBO are compared with ordinary TLBO, differential evolution (DE), particle swarm optimization (PSO) and linear programming (LP). It was observed that ETLBO performed better in terms of better global searchability and faster convergence than TLBO, DE, PSO and LP.

Keywords Optimization · Reservoir operation · Teaching learning-based optimization · Elitist algorithm · Ukai dam

16.1 Introduction

Complexities in water allocation due to increasing demands and shortage of water resources have increased the importance of obtaining the sustainable development and maximizing benefits (Bahrami et al. 2018). Due to the disparity between the demand and supply, it causes pressure on water supply, conflict among other water users, competition and impact on the environment (Chang and Wang 2013). It is a challenging and important for the policymaker to make effective water allocation and water management issues. In order to solve these problems, optimization of

V. Kumar (✉)
Civil Engineering Department, G. H. Raisoni College of Engineering and Management, Pune, Maharashtra, India

S. M. Yadav
Civil Engineering Department, Sardar Vallabhbhai National Institute of Technology, Surat, Gujarat, India

R. Jha et al. (eds.), *Water Resources Management and Reservoir Operation*, Water Science and Technology Library 107,
https://doi.org/10.1007/978-3-030-79400-2_16

an existing project is an alternative solution (Kumar and Yadav 2020a). Either traditional or evolutionary-based algorithm can be applied to solve the reservoir operation problems. The traditional techniques used in reservoir operation are linear programming (LP), nonlinear programming (NLP) and dynamic programming (DP). Despite the effectiveness of these methods, it has its own limitation. Such as LP need linear objective function and constraints, DP is cursed by dimensionality and NLP is not able to solve the non-convex problem effectively (Kumar and Yadav 2020b). These are the few limitations of the traditional methods, apart, it is observed to stick in local optimum solution.

An evolutionary algorithm is free from these limitations and able to provide a global solution. Over the past few decades, evolutionary algorithms have gained interest among researchers. Different algorithms have been used to solve the reservoir operation problem, for example, particle swarm optimization (PSO) (Kumar and Reddy 2007; Afshar 2013; Bai et al. 2017), genetic algorithm (GA) (Chang et al. 2010), differential evolutionary (DE) (Regulwar et al. 2010), weed optimization algorithm (WOA) (Asgari et al. 2016), cuckoo search (CS) (Ming et al. 2015), artificial bee colony (ABC) (Hossain and El-shafie 2014) and shark algorithm (Ehteram et al. 2017). The existing algorithms are having few limitations. Evolutionary algorithms are having two type of parameters, first, common controlling parameters, i.e. population size and a number of iterations, and second is individual algorithm-specific parameters, such as PSO used internal parameters like social and cognitive parameters and inertia weight. GA required mutation parameters, crossover parameters and reproduction parameters. DE used crossover rate and scaling factor. ABC necessitates the number of employed bees, scout bees and onlookers. These parameters are called algorithm-specific parameters, and it is different for different algorithms. Before executing the algorithm, it is required to tune these parameters, improper selection of these parameters affect the overall performance of the algorithm. Apart, other algorithms such as cuckoo search, bat algorithm, ant colony optimization, weed optimization, etc., used in reservoir operations have their own parameters.

Teaching learning-based optimization (TLBO) is a newly developed algorithm developed based on human behaviour of teaching and learning process (Rao et al. 2011). TLBO do not have any algorithm-specific parameters. Rao et al. (2012) used TLBO to optimize the continuous nonlinear problem, and it performed better than other algorithms. Baykasoğlu et al. (2014) used TLBO for flow shop and job shop scheduling problem and was found to have considerable potential compared to other algorithms. Bayram et al. (2015) applied TLBO over dissolved oxygen problem. Baghlani et al. (2017) used TLBO to solve steel truss sizing optimization problem. Kumar and Yadav (2018) used TLBO to solve a multi-reservoir operation problem. Kumar and Yadav (2019) used ETLBO to optimize the cropping pattern over two different models and compared the results with the elitist jaya algorithm.

This paper presents the improved version of TLBO, by introducing the elitist concept in order to improve the convergence, faster evolutions process, and global search ability. The objective of the study is to maximize the water allocation for Ukai reservoir to supply water for irrigation, domestic and industrial uses at different dependable inflow. Elitist teaching learning-based optimization (ETLBO) has been used to optimize water allocation, for four different models with different dependable inflow, namely 60, 65, 70 and 75%. The results from ETLBO are compared with ordinary TLBO, DE, PSO and LP. The following section describes methods and materials.

16.2 Methods and Materials

16.2.1 Differential Evolution (DE)

DE was proposed by (Storn and Price 1996) as an effective heuristic technique used for solving optimization problems. DE optimized the problem based on a set of population and updating the population with existing ones according to the simple formula. And, then keeping the best solution out of the two which score better functional value or fitness to hand the optimization problem. The algorithm follows the following steps.

Step 1: Decide the parameters and generate the initial population vector solution, i.e. $Y^j_{k,G} = \{y^1_{k,G}, y^2_{k,G}, \ldots, y^D_{k,G}\}$ using the expression as per Eq. (16.1).

$$y^j_{k,0} = y^j_{\min} + R_{k,j} * \left(y^j_{\max} - y^j_{\min}\right) \tag{16.1}$$

where $y^j_{k,0}$ is the initial population at kth component for jth vector at generation $G = 0$, $k = 1, 2, \ldots, n$, n is the population size, $j = 1, 2, \ldots, p$, p is dimensional parameter vector, $y^j_{\min}$ and $y^j_{\max}$ are the lower and upper limit, $R_{k,j}$ is the random number.

Step 2: Produce mutation vector, i.e. $V^j_{k,G} = \{V^1_{k,G}, {}^2_{k,G}, \ldots, V^D_{i,G}\}$ for each target vector $Y^j_{k,G}$, mutation operator is used.

$$V^j_{k,G} = Y_{r^k_1,\,G} + F * \left(Y_{r^k_2,\,G} - Y_{r^k_3,\,G}\right) \tag{16.2}$$

where r^k_1, r^k_2, and r^k_3 are the integers randomly generated between [1, NP], F is the scaling factor parameter varies between [0, 2].

Step 3: Next crossover operation is applied wherein each target vector $Y^j_{k,G}$ and its corresponding mutation vector $V^j_{k,G}$ to obtained a trail vector $U^j_{k,G} = \{U^1_{k,G}, U^2_{k,G}, \ldots, U^D_{k,G}\}$. The basic binomial crossover is defined as Eq. (16.3).

$$U^j_{k,G} = \begin{cases} V^j_{k,G}, & \text{if (rand[0, 1]} \le \text{CR) or } (j = j_{\text{rand}}) \\ Y^j_{k,G}, & \text{otherwise} \end{cases} \tag{16.3}$$

where j_{rand} is the integers randomly generated between [1, NP], CR is the crossover rate parameter varies between [0, 1].

Step 4: Selection operation

$$Y^{j}_{k,G+1} = \begin{cases} U^{j}_{k,G}, & \text{iff}\left(U^{j}_{k,G}\right) \leq f\left(Y^{j}_{k,G}\right) \\ Y^{j}_{k,G}, & \text{otherwise} \end{cases} \quad (16.4)$$

Step 5: Increase the generation to $G = G + 1$. The algorithm reiterates until the maximum iterations are accomplice.

16.2.2 Particle Swarm Optimization (PSO)

PSO was proposed by Eberhart and Kennedy (1995). It is a metaheuristic-based swarm intelligence established optimization technique, similar to other population-based algorithm. PSO works similar to the natural grouping as birds flock and fish school. It provides the optimal solution based on the individual and social behaviour (Shourian et al. 2008). The algorithm follows the following steps.

Step 1: Decide the parameters and generate the initial group of random particles, for each particle, generate positions Y_k and velocity v_k.

Step 2: Evaluate the fitness and obtained $Pbest_k$, i.e.

$$Pbest_k = \begin{cases} Pbest_k(i-1), & \text{if } F(Pbest_k(i)) \leq F(Pbest_k(i-1)) \\ Pbest_k(i), & \text{if } F(Pbest_k(i)) > F(Pbest_k(i-1)) \end{cases} \quad (16.5)$$

where $Pbest_k$ represents the best position of the individual, I represent the iteration, i.e. $I = 1, 2, \ldots, i$, of the kth individual.

Step 3: Next is find the global best position from the swarm, i.e.

$$Gbest_k(I) = \max\{F(Gbest_1(i)), F(Gbest_2(i)), \ldots, F(Gbest_n(i))\} \quad (16.6)$$

Step 4: Update the velocity using Eq. (16.7) if $v_k > v_{\max}$ then $v_k = v_{\max}$; if $v_k < v_{\min}$ then $v_k = v_{\min}$.

$$\begin{aligned} v_k(i+1) = w(i) * v_k(i) + c_1 * r_1(i) * (Pbest_k(i) - Y_k(i)) \\ + c_2 * r_2(i) * (Gbest_k(i) - Y_k(i)) \end{aligned} \quad (16.7)$$

where v_k and Y_k are the velocity and position of kth individual, respectively. The internal parameters are $w(i)$, c_1 and c_2, term $w(i)$ represent the internal weight, c_1 is the cognitive parameters and c_2 is the social parameters.

Step 5: Update the velocity using Eq. (16.8).

$$Y_k(i+1) = Y_k(i) + v_k(i+1) \tag{16.8}$$

This complete one cycle, the algorithm reiterates until the maximum iterations are accomplice.

16.2.3 Teaching Learning-Based Optimization (TLBO)

TLBO is a metaheuristic optimization technique similar to other population-based algorithm proposed by Rao et al. (2011). It works on the human behaviour of teaching and learning process. It consists of a two-phases teaching and learning. The algorithm follows the following steps.

Step 1: Decide the common controlling parameters and generate the initial population random between the bounds, using Eq. (16.9).

$$\text{Randomly Generated Population } y_k = y_{k,\min} + R_k * \left(y_{k,\max} - y_{k,\min}\right) \tag{16.9}$$

where y_k is the initial population at kth component, $k = 1, 2, \ldots, n$, n is the population size, $y_{k,\min}$ and $y_{k,\max}$ are the lower and upper limit, R_k is the random number between [0, 1].

Step 2: From the randomly generated population identify the best solution.

Step 3: Teaching phase, wherein teacher teaches to improve the results, as presented in Eqs. (16.10) and (16.11).

$$\text{Difference mean} = r * (y_{\text{best}} - y_{\text{mean}}) \tag{16.10}$$

$$y_{\text{new}} = (y_{\text{old}} + \text{Difference mean}) \tag{16.11}$$

where y_{best} and y_{mean} are the best and mean solution, y_{new} and y_{old} are the modified and old solution. Better functional value obtained by y_{new} and y_{old} is selected at the end of the teaching phase.

Step 4: Learning phase, wherein any two learners tries to exchange their knowledge for better results. For example, any two random learners, i.e. y_1 and y_2, where $1 \neq 2$

$$\text{iff}(y_1) < f(y_2);\ \text{Then, } y_{\text{new},\,1} = y_{\text{old},1} + r_i * (y_2 - y_1) \tag{16.12}$$

$$\text{iff}(y_1) > f(y_2);\ \text{Then, } y_{\text{new},\,1} = y_{\text{old},1} + r_i * (y_1 - y_2) \tag{16.13}$$

Again, the better functional value is selected between the old and new solution.

Step 5: This complete one cycle, the algorithm reiterates until the maximum iterations are accomplice.

16.2.4 *Elitist Teaching Learning-Based Optimization (ETLBO)*

ETLBO is the improved version of TLBO, wherein elitist concept has been applied with ordinary TLBO. The elitist concept has been applied in the different evolutionary algorithm; the same concept is applied in TLBO, wherein it preserves the best solution. Elitism helps to remove the worst solution with the best elite solution; the replacement depends on the number of elite sizes. The working steps are similar with TLBO up to learning phase. At the end of the learning phase, an elite solution is identified. The greedy selection process is adapted to update the solution. This complete one cycle, the algorithm reiterates until the maximum iterations are accomplice.

16.3 Study Area and Data Collection

Tapi River is an interstate River of India originates from Betul district at an elevation of 752 m above MSL and having a total length of 724 km. It drains an area of 65,145 km^2 into Gulf of Khambhat, Arabian Sea. Ukai dam was constructed across the Tapi River in 1972. It was constructed for multipurpose such as irrigation, water supply, power generation and flood control. It is having a total capacity of 8510 million cubic metres (MCM), which include live storage of 7079.19 MCM and dead storage of 1430.82 MCM. Table 16.1 shows the salient features of the Ukai reservoir. The reservoir water is allocated for domestic purpose, industrial purpose and irrigation purpose. Downstream of the Ukai dam is Surat city having a population of approximately 6.6 million. Figure 16.1 shows the index map of the study area,

Table 16.1 Salient features of Ukai reservoir

General ıtem	Details
Mean annual rainfall in the watershed	785 mm
Maximum annual rainfall in the watershed	1191 mm
Coordinates	21° 14′ 53.52″ N 73° 35′ 21.84″ E
Catchment area	62,225 Km^2
Top of dam	111.252 m
Area	65,145 Km^2
State in the basin	Maharashtra, Madhya Pradesh, and Gujarat
Road width on the spillway	6.706 m
Type of spillway	Radial

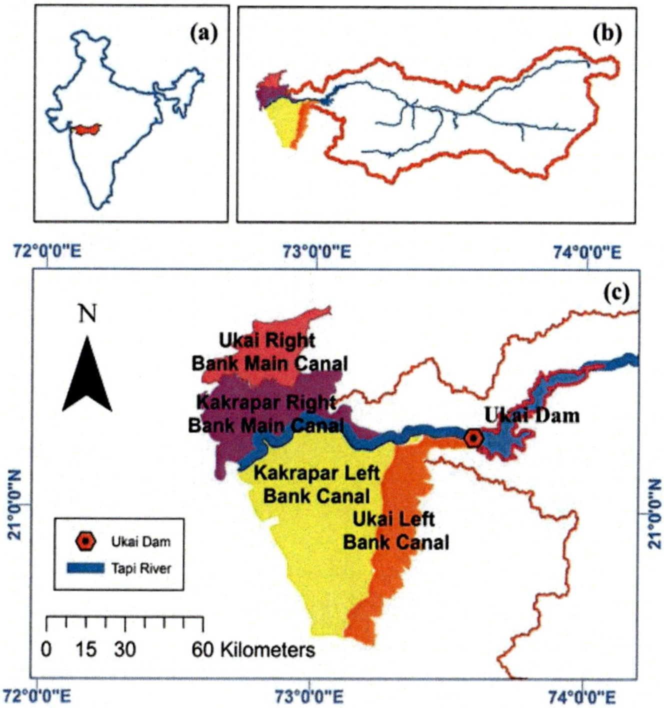

Fig. 16.1 Index map **a** India map showing Tapi basin, **b** Tapi basin and **c** Cultivated land from Ukai reservoir

wherein Fig. 16.1a represents the India map showing Tapi basin, Fig. 16.1b represents the Tapi catchment area, and Fig. 16.1c shows the location of the dam and cultivated land from Ukai reservoir. Four different canal networks are present in different colour combinations, namely Ukai right and left bank main canal and Kakrapar right and left bank main canal, which was used for irrigation. Figure 16.2 shows the line diagram of Ukai dam, and it represents the various demands and losses.

The data required for the present study are water allocation for irrigation, domestic demands, industrial demands, storage of dam, the inflow of the reservoir and evaporation losses. The data were collected from Surat Irrigation Circle (SIC) and Ukai Left Bank Division (ULBD).

16.4 Mathematical Models

The objective function and the constraints are listed as follows.

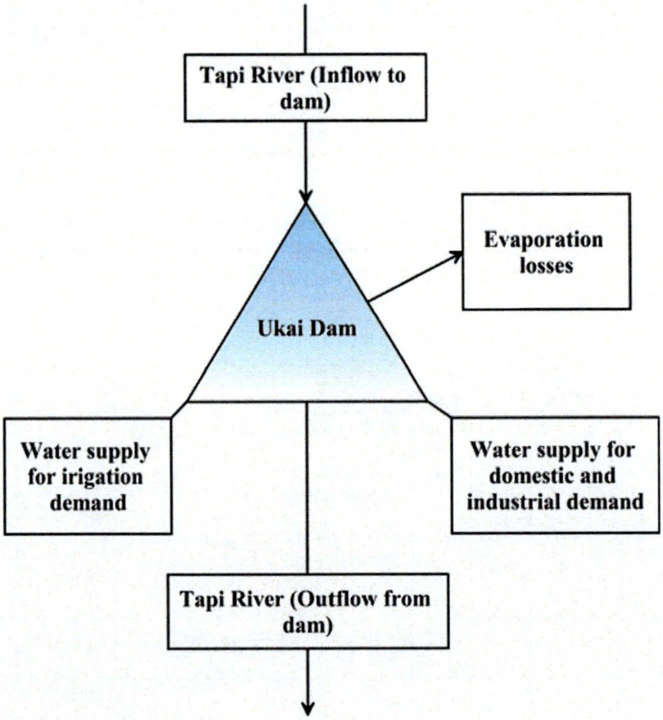

Fig. 16.2 Line diagram of Ukai dam

16.4.1 Objective Function

The aim of the present study is to maximize the water allocation for irrigation demand, domestic demand and industrial demands, over a year from Ukai reservoir, in consideration of hydropower benefits. The objective function is mathematically expressed as Eq. (16.14).

$$\text{Max } F(B) = \sum_{t=1}^{12} R_{i,t} + \sum_{t=1}^{12} R_{d,t} + \sum_{t=1}^{12} R_{in,t} \tag{16.14}$$

where $F(B)$ is the maximum water allocated for different purpose, $R_{i,t}$ is the water allocated for irrigation purpose in MCM, $R_{d,t}$ is the water allocated for domestic purpose in MCM, $R_{in,t}$ is the water allocated for industrial purposein MCM and t is the time period, i.e. $t = 1, 2, \ldots, 12$.

To satisfied reservoir constraints, the penalty is applied to the reservoir. The objective function is modified as per Eq. (16.15).

$$\text{Max } F(B) = \sum_{t=1}^{12} R_{i,t} + \sum_{t=1}^{12} R_{d,t} + \sum_{t=1}^{12} R_{in,t} - p(n) * \text{abs}\left(p_{(n,t)}\right)^2 \tag{16.15}$$

where $p(n)$ is the penalty parameters $p(n) = 5$, and $p_{(n,t)}$ is the penalty function.

16.4.2 Constraints

Following are constraints that are subjected to the reservoir.

Storage capacity constraint

Monthly storage of the reservoir should not be less than the minimum monthly capacity at which hydropower is generated and should not be more than the maximum monthly capacity, for time t month, i.e. $t = 1, 2, \ldots, 12$.

$$S_{t,\text{Min}} \leq S_t \leq S_{t,s\text{Max}} \tag{16.16}$$

where $S_{t,\min}$ is the monthly minimum storage capacity of the reservoir in MCM, $S_{t,\max}$ is the monthly maximum storage capacity of the reservoir in MCM and S_t is the monthly storage in MCM.

Reservoir storage continuity constraint

The reservoir storage is governed by the continuity equation, represented by Eq. (16.17).

$$S_{t+1} = S_t + I_t - \left(R_{i,t} + R_{d,t} + R_{in,t}\right) - E_t \tag{16.17}$$

Where S_t is the active storage during month t, I_t is the reservoir inflow during month t in MCM, E_t is the evaporation loss during month t in MCM and S_{t+1} is the storage during month $(t + 1)$.

Irrigation demand constraint

Monthly irrigation demand should be equal to or less than the maximum water allocated for irrigation and it should be equal to or more than the minimum water allocated for irrigation.

$$R_{i,\min,t} \leq R_{i,\ t} \leq R_{i,\max,t} \tag{16.18}$$

Where $R_{i,\min,t}$ is the minimum water allocated for irrigation and $R_{i,\max,t}$ is the maximum water allocation for irrigation, during month t in MCM.

Domestic water demand constraint

Monthly domestic water demand should be equal to or less than the maximum water allocated for domestic demand, and it should be equal to or more than the minimum water allocated for domestic demand.

$$D_{d,\min,t} \leq R_{d,t} \leq D_{d,\max,t} \tag{16.19}$$

where $D_{d,\min,t}$ is the minimum water allocated for domestic demand and $D_{d,\max,t}$ is the maximum water allocation for domestic demand, during month t in MCM.

Industrial water demand constraint

Monthly industrial water demand should be equal to or less than the maximum water allocated for industrial demand, and it should be equal to or more than the minimum water allocated for industrial demand.

$$ID_{in,\min,t} \leq R_{in,t} \leq ID_{in,\max,t} \tag{16.20}$$

where $ID_{in,\min,t}$ is the minimum water allocated for industrial demand and $ID_{in,\max,t}$ is the maximum water allocation for industrial demand, during month t in MCM.

Non-negativity constraint

$$R_{i,t} > 0, \ R_{d,t} > 0, \ R_{in,t} > 0 \tag{16.21}$$

16.5 Results and Discussion

The aim was to maximize the water allocation from Ukai reservoir, at various dependable inflows. The Weibull method was used to find the different dependable inflow (Subramanya 2013). The different population sizes were used, i.e. 10, 20, 40, 60, 80, 100 and 150 and the iteration size as 50,000. The internal parameters in DE, i.e. the lower and upper scaling factors were taken as 0.2 and 0.8, respectively, and the crossover probability was taken as 0.2. The internal parameters in PSO, i.e. the social and cognitive parameters were taken as 2.0 and 1.5, respectively, and inertia weight as 1.

Table 16.2 shows the comparative results obtained using ETLBO, TLBO, PSO and DE, in the form best, worst and average solution. The results were averaged over 15 different runs. A very small standard deviation and coefficient of variation have been observed for all the algorithms. Table 16.2a shows the best results over 15 different runs, wherein ETLBO, TLBO and PSO were having same results, and DE and LP results were almost similar but less than other three algorithms. Table 16.2b shows the average results obtained over 15 different runs, and it was found

Table 16.2 Comparative results (a) best, (b) average and (c) worst, at various dependable inflows

Results	Algorithm	Maximum allocation in MCM			
		60%	65%	70%	75%
(a) Best	ETLBO	3224.520	4023.190	4672.770	5351.020
	TLBO	3224.520	4023.190	4672.770	5351.020
	PSO	3224.520	4023.190	4672.770	5351.020
	DE	3219.339	4022.210	4672.160	5350.420
	LP	3223.920	4022.579	4672.170	5350.420
(b) Average	ETLBO	3224.520	4023.190	4672.770	5351.020
	TLBO	3224.515	4023.187	4672.764	5351.018
	PSO	3224.513	4023.183	4672.747	5351.014
	DE	3219.249	4022.149	4672.123	5350.331
(c) Worst	ETLBO	3224.520	4023.190	4672.770	5351.020
	TLBO	3224.490	4023.160	4672.730	5351.010
	PSO	3224.480	4023.150	4672.600	5350.980
	DE	3219.039	4022.010	4672.000	5350.130

Bold value shows the best solution for (a), (b) and (c) separately

that ETLBO results were better as compared to other algorithms. Table 16.2c shows the worst results obtained over 15 different runs, and it was observed that ETLBO results were better as compared to other algorithms. Overall it was found that ETLBO performed better when compared to TLBO, PSO, DE in term of average results and worst solution. The global optimum solution obtained using ETLBO for 60%, 65%, 70% and 75% dependable inflow were 3224.520, 4023.190, 4672.770 and 5351.020, respectively, in MCM, and similar were obtained using TLBO and PSO. But it was found that ETLBO had better uniformity in the solution as compared to TLBO and PSO. DE and LP results were inferior than other algorithms. Figure 16.3a–d shows the convergence curve for water allocated having 60, 65, 70 and 75% dependence flow. It was found that ETLBO had better convergence rate as compared to other algorithm in all the dependable inflow. Note, iteration up to 5000 is only shown.

Next, the best optimum solution obtained from ETLBO at different dependable inflow water is compared with the actual water releases. Figure 16.4 shows the comparison between the actual and computed water releases. Figure 16.4, it was observed that at 60% dependable inflow much variations were observed with actual releases of water than other dependable inflow. At 75%, least variation has been observed as compared with actual releases of water.

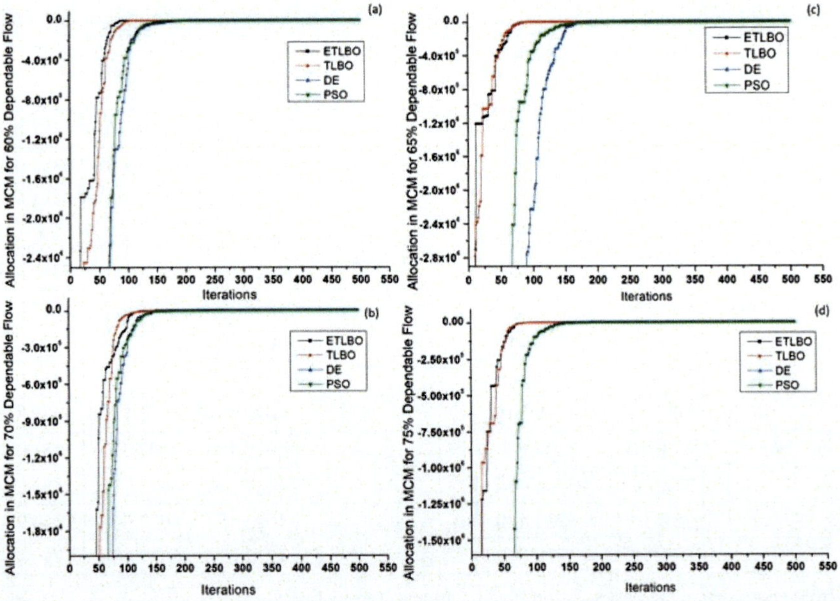

Fig. 16.3 Convergence curve (**a**) 60% (**b**) 65% (**c**) 70% (**d**) 75% dependable inflow

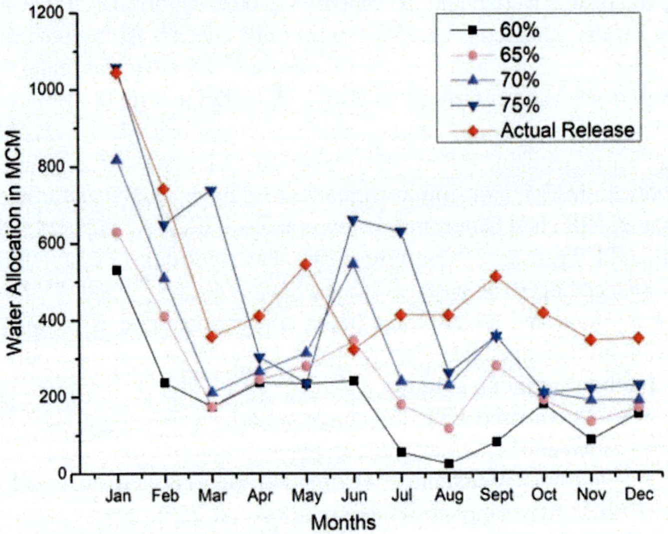

Fig. 16.4 Comparison of actual and computed release for different dependable inflow

16.6 Conclusion

The present study introduces an improved version of TLBO, by introducing the elitist concept. The objective was to maximize the water allocation from Ukai reservoir to supply water for various needs at different dependable inflow. ETLBO and TLBO were chosen because it does not require internal parameters. To turn the algorithm as it is commonly observed in other algorithm, it saves a lot of time. The results from ETLBO were compared with ordinary TLBO, PSO, DE and LP. It was found that ETLBO performed better when results are compared as average and worst solution. ETLBO results were uniform, faster convergence rate and better global search ability. When comparison was done with actual releases, it was found that 75% dependable inflow variations were less compared with other dependable inflows. Thus, it is suggested to operate the reservoir as per the 75% dependable inflow level model.

References

Afshar MH (2013) Extension of the constrained particle swarm optimization algorithm to optimal operation of multi-reservoirs system. Int J Electr Power Energy Syst 51:71–81

Asgari HR, Bozorg Haddad O, Pazoki M, Loáiciga HA (2016) Weed optimization algorithm for optimal reservoir operation. J Irrig Drain Eng 142(2):04015055

Baghlani A, Makiabadi MH, Maheri MR (2017) Sizing optimization of truss structures by an efficient constraint-handling strategy in TLBO. J Comput Civ Eng 31(4):04017004

Bahrami M, Bozorg-haddad O, Chu XF (2018) Application of Cat Swarm Optimization Algorithm for Optimal Reservoir Operation. J Irrig Drain Eng 144(1):04017057. https://doi.org/10.1061/(ASCE)IR.1943-4774.0001256

Bai T, Kan Y, Chang J, Huang Q, Chang FJ (2017) Fusing feasible search space into PSO for multi-objective cascade reservoir optimization. Appl Soft Comput 51:328–340

Baykasoğlu A, Hamzadayi A, Köse SY (2014) Testing the performance of teaching-learning based optimization (TLBO) algorithm on combinatorial problems: flow shop and job shop scheduling cases. Inf Sci 276:204–218

Bayram A, Uzlu E, Kankal M, Dede T (2015) Modeling stream dissolved oxygen concentration using teaching–learning based optimization algorithm. Environ Earth Sci 73(10):6565–6576

Chang FJ, Wang KW (2013) A systematical water allocation scheme for drought mitigation. J Hydrol 507:124–133

Chang LC, Chang FJ, Wang KW, Dai SY (2010) Constrained genetic algorithms for optimizing multi-use reservoir operation. J Hydrol 390(1–2):66–74

Eberhart R, Kennedy J (1995) A new optimizer using particle swarm theory. In: MHS'95. Proceedings of the sixth international symposium on micro machine and human science. IEEE, pp 39–43

Ehteram M, Karami H, Mousavi SF, El-Shafie A, Amini Z (2017) Optimizing dam and reservoirs operation based model utilizing shark algorithm approach. Knowl -Based Syst 122:26–38

Hossain MS, El-shafie A (2014) Performance analysis of artificial bee colony (ABC) algorithm in optimizing release policy of Aswan high dam. Neural Comput Appl 24(5):199–1206

Kumar ND, Reddy JM (2007) Multipurpose Reservoir Operation Using Particle Swarm Optimization. J Water Resour Plan Manag 133(3):192–201

Kumar V, Yadav SM (2018) Optimization of reservoir operation with a new approach in evolutionary computation using TLBO algorithm and Jaya algorithm. Water Resour Manag 32(13):4375–4391

Kumar V, Yadav SM (2019) Optimization of cropping patterns using Elitist-Jaya and Elitist-TLBO algorithms. Water Resour Manage 33(5):1817–1833. https://doi.org/10.1007/s11269-019-02204-z

Kumar V, Yadav SM (2020a) Self-adaptive multi-population based Jaya algorithm to optimize the cropping pattern under a constraint environment. J Hydroinformatics 1–18. https://doi.org/10.2166/hydro.2019.087

Kumar V, Yadav SM (2020b) Optimization of water releases from Ukai reservoir using Jaya algorithm. Advances in intelligent systems and computing, vol 949. Springer, Singapore. https://doi.org/10.1007/978-981-13-8196-6_29

Ming B, Chang JX, Huang Q, Wang YM, Huang SZ (2015) Optimal operation of multi-reservoir system based-on cuckoo search algorithm. Water Resour Manag 29(15):5671–5687

Rao RV, Savsani VJ, Vakharia DP (2011) Teaching-learning-based optimization: a novel method for constrained mechanical design optimization problems. CAD Comput Aided Des 43(3):303–315

Rao RV, Savsani VJ, Vakharia DP (2012) Teaching-learning-based optimization: an optimization method for continuous non-linear large scale problems. Inf Sci 183(1):1–15

Regulwar DG, Choudhari SA, Raj PA (2010) Differential evolution algorithm with application to optimal operation of multipurpose reservoir. J Water Resour Prot 02(06):560–568

Shourian M, Mousavi SJ, Tahershamsi A (2008) Basin-wide water resources planning by integrating PSO algorithm and MODSIM. Water Resour Manag 22(10):1347–1366

Storn R, Price K (1996) Minimizing the real functions of the ICEC'96 contest by differential evolution. In: Proceedings of IEEE international conference on evolutionary computation. IEEE, pp 842–844

Subramanya K (2013) Engineering hydrology. Tata McGraw-Hill Educ 45–60

Chapter 17
Prediction of Reservoir Submerged Sediment Density

Y. C. Jabbar and S. M. Yadav

Abstract This study proposes a density prediction model for reservoir sediment deposition using artificial neural networks (ANNs). To compute the reservoir capacity loss, it is necessary to estimate the weight per unit submerged sediment volume as the sediment transported in a river is measured in gravimetric terms as sediment load. Understanding the sensitivity of the estimated density to relate catchment sediment yield with the reservoir deposition rate, ANNs are utilized to precisely compute submerged sediment density. A dataset with 262 field observed densities for the reservoirs which always remain submerged is prepared and used. Three input variables, sand, silt, and clay proportion are selected. Then, the method of training and validation of the ANNs for the density prediction is presented. The model results show that the ANN model is flexible and robust to capture the complex physical process and is better than the Lara Pemberton empirical relationship. The proposed trained network, having the best predictive capability, is given as a MATLAB code.

Keywords Sediment volume · Reservoir · Deposition · Density · ANN

17.1 Introduction

To determine the economic life of a dam in a sediment-bearing river, the most important challenge is to estimate the dead storage space. The dead storage allocation is done based on the estimated sediment deposition in the course of the useful life of a reservoir. The gradual reduction in reservoir capacity, with time, depends on the sediment load trapped by the reservoir and on the density of deposition (Jabbar and Yadav 2019a, b). Sediment load can be precisely estimated in gravimetric terms from the hydrometric observations, whereas the density of deposited sediments is variable

Y. C. Jabbar (✉)
Department of Civil Engineering, Sitarambhai Naranji Patel Institute of Technology & Research Centre, Umrakh, Surat, Gujarat 394601, India

S. M. Yadav
Department of Civil Engineering, S. V. National Institute of Technology, Surat, Gujarat 395007, India

R. Jha et al. (eds.), *Water Resources Management and Reservoir Operation*,
Water Science and Technology Library 107,
https://doi.org/10.1007/978-3-030-79400-2_17

in the areal extent (of the reservoir submergence-area) and increases with the depth of deposition. The submerged density of the deposited sediment may range from 363 to 2000 kg/m^3 (Strand and Pemberton 1982).

In an existing built dam-reservoir, the sediment distribution and subsequent variation of the density can be studied by sampling the sediments. Point sampling represents the density of the sediment deposited at a point. In practice, however, capacity loss determination requires the knowledge of bulk-density over the submerged area of reservoir (Jabbar and Yadav 2019a, b). To convert the point sampled density into a reservoir-representative bulk-density, huge number of sampling points should be established. Such multi-point samplings are not feasible in large reservoirs, as the reservoir can be deep up to hundreds of meters (Annandale et al. 2016). Estimating a value of density from secondary data is carried out to predict the reservoir sedimentation and capacity loss from the hydrometric observations (Bussi et al. 2013; Marineau and Wright 2017). Sand, silt, and clay content of sediments are associated with the initial weight per unit volume for different operational characteristics of the reservoirs.

Lane and Koelzer (1943) first highlighted the importance of the density of sediments deposited in reservoirs. They have analyzed the data from different rivers and reservoirs and developed a relationship to predict initial density. Considering the effect of consolidation of clay and silt for a period of T-years, the density computation was proposed. For sediment composed of sand, silt, and clay material, the unit weight for each material is multiplied to their respective fraction (Eq. 17.1).

$$W_i = \left[\left(W_c\frac{P_c}{100} + k\log_{10}T\right) + \left(W_m\frac{P_m}{100} + k\log_{10}T\right) + \left(W_s\frac{P_s}{100} + k\log_{10}T\right)\right] \tag{17.1}$$

where, W_i = Initial density in kilograms per cubic meter.

K_c, K_m and K_s are constants of clay, silt, and sand, respectively, based on type of reservoir operation.

P_c, P_m and P_s are percentages of clay, silt, and sand, respectively, of the incoming sediment.

W_c, W_m and W_S are coefficients of clay, silt, and sand, respectively of the incoming sediment as given in Table 17.1. Thus, initial density, i.e., density of a year of deposited sediment can be expressed as Eq. 17.2.

$$W_i = W_c\frac{P_c}{100} + W_m\frac{P_m}{100} + W_s\frac{P_s}{100} \tag{17.2}$$

Lara and Pemberton (1965) modified the coefficients of density (W_s, W_m and W_c) with the use of additional data on the density of deposited material from reservoir surveys (Table 17.2). They have recommended Miller (1953) consolidation equation to compute density for a period of T-years. As a higher number of data is utilized to

Table 17.1 Initial weight as per Lane and Koelzer (1943) for different reservoir operations

Reservoir operation	W_c (kg/m^3)	K_c	W_m (kg/m^3)	K_m	W_s (kg/m^3)	K_s
Type-1[a]	480	256.3	1041	91.3	1490	0
Type-2[b]	787	171.4	1185	43.3	1490	0
Type-3[c]	961	96.1	1265	16.0	1490	0
Type-4[d]	1250	0	1314	0	1490	0

Note The values under the table are originally (Lane and Koelzer 1943) in the lbs./cu. Ft. which are converted to kg/m^3

[a]Sediment always submerged or nearly submerged (type-1)

[b]Normally moderate to considerable reservoir drawdown (type-2)

[c]Reservoir normally empty (type-3)

[d]River bed sediments (type-4)

Table 17.2 Lara and Pemberton (1965) modified the initial weight for different reservoir operations

Reservoir operation	W_c (kg/m^3)	W_m (kg/m^3)	W_s (kg/m^3)
Type-1[a]	416	1121	1554
Type-2[b]	561	1137	1554
Type-3[c]	641	1153	1554
Type-4[d]	961	1169	1554

[a]Sediment always submerged or nearly submerged (type-1)

[b]Normally moderate to considerable reservoir drawdown (type-2)

[c]Reservoir normally empty (type-3)

[d]River bed sediments (type-4)

obtain modified initial weight coefficients (Table 17.2), Lara and Pemberton (1965) advised the use of the modified coefficients for the determination of initial weight values. No further modification in Lara and Pemberton (1965) initial weight coefficients has been done, and the same are used up to date to determine the initial unit weight (Marineau and Wright 2017). With the complex relationship between the sediment size fraction and initial density, empirically derived (regression) relationships have shortcomings in depicting the nonlinearity. In addition to it, the proportion of sediments are interrelated to each other (multicollinearity). Thus, identifying a relationship from them using the regression approach is not advised (Donald and Glauber 1967). The fraction of sand, silt, and clay has to add up to be one (100%); an increase in the proportion of one may consequently decrease the other two fractions of sediments. This is problematic; it can be hard to unravel which of the sediment fraction best describes its shared variance with density. On the other hand, artificial neural networks (ANNs) have proven fit to model the behavior of extremely complex processes in numerous fields (Yegnanarayana 2009).

In the present work, an attempt is made to model the impact of particle size fraction on the initial density of deposition for the sediments, which always remain submerged. Besides, no study exists in the literature yet to model the density estimates using ANN techniques. After several trials with field data, the architecture of ANN model is determined. The network training is performed by the inputs of particle size fractions (sand, silt and clay) and targets as observed/sampled densities of different reservoirs.

17.2 Materials and Methods

17.2.1 Data

In this study, field data published by Lara and Pemberton (1963) are used. These data were collected from different reservoirs having sediment always remained submerged. The statistics of the sampled density is presented in Table 17.3. It is observed from the sampled density statistics that the standard deviation of the density observed is between 35 and 493 kg/m^3. The standard deviation of sand, silt, and clay was observed to be 3.19, 3.32, and 1.30, respectively, for the reservoir with a minimum standard deviation of observed density.

17.2.2 ANN Model Design

A neural network has proved itself as an elite soft computing tool and can be applied to analyze, predict, or forecast the problems pertaining to curve fitting (Liu et al. 2013; Noori and Kalin 2016; Safari et al. 2016; Yang et al. 2009), pattern recognition, clustering, and dynamic time series (Raman and Sunilkumar 1995). With the advancement of computing power, deep and shallow neural networks have gained importance in the field of mapping complex and nonlinear relationships. Curve fitting problem is addressed in the present study by ANN using MATLAB scripts. Sand, silt, and clay proportion of the reservoir deposited sediments are given as inputs to estimate the output of field density. The network is developed to predict the deposited sediments density for the type of reservoir in which sediment always remain submerged. A shallow feed-forward neural network of two layers, consisting hidden, and output layer is designed (Basheer and Hajmeer 2000). If data grouping is not done correctly, the network may result in a very rigid memory-based network (Bowden et al. 2002; Maier and Dandy 2000). The efficiency of the network to map complex nonlinear relationships is directly influenced by the number of nodes selected in network design (Tokar and Johnson 1999). Thus, neural networks are designed in the present study by changing two parameters. One, grouping of data in different proportions for training, calibration, and validation. Second, number of nodes in the hidden layer is varied. The network is trained with Levenberg–Marquardt backpropagation algorithm. The transfer function computes the net layer output from its net input. Without the use of a transfer function, the neural network may merely reduce to a group of linear

Table 17.3 Statistics of the sampled density observations

Sr. No	Reservoir	Location	Surveyed by	Sample no	Observed density statistics				Average of		
					Max	Min	Average	Standard deviation	Clay	Silt	Sand
1	Harry strunk	Nebraska	USBR[a]	22	1590.63	480.55	1144.45	339.31	14.39	61.81	22.67
2	Leavenworth country state lake	Kansas	SCS[b]	9	1169.35	592.68	923.73	213.11	35.18	61.71	3.11
3	Lake Calhoun	Illinois	SCS	2	1089.26	760.88	925.07	232.20	32.15	67.60	0.25
4	Lake Bracken	Illinois	SCS	3	776.90	584.67	664.23	100.30	41.90	57.30	0.80
5	Lanecaster reservoir	South Carolina	SCS	7	1263.86	631.13	1011.45	209.90	28.77	62.69	8.54
6	Conchas reservoir	New Mexco	CE[c]	8	1345.55	331.58	755.67	334.46	39.63	47.25	13.13
7	Tarryall reservoir	Colorado	FS[d]	3	1400.01	499.78	1066.30	493.21	8.53	32.67	58.63
8	Lake McBride	Iowa	SCS	5	993.14	897.03	948.29	34.73	17.80	80.00	2.20
9	Lake Olathe	Kansas	SCS	1	961.11	961.11	961.11	–	22.40	22.10	55.50
10	Lake EI Darado	Kansas	SCS	1	1057.22	1057.22	1057.22	–	15.20	27.10	57.70
11	Fort supply reservoir	Oklahoma	CE	26	1920.61	488.56	990.80	472.93	47.85	23.19	29.35
12	Lake Texoma	Oklahoma and Texas	CE	59	1729.99	560.65	1197.31	316.21	24.76	41.92	33.02
13	Lake Bloomington	Illinois	ISWS[e]	16	1349.40	439.87	739.05	263.26	81.56	13.81	4.63

(continued)

Table 17.3 (continued)

Sr. No	Reservoir	Location	Surveyed by	Sample no	Observed density statistics				Average of		
					Max	Min	Average	Standard deviation	Clay	Silt	Sand
14	Lake Decatur	Illinois	SCS	58	1129.46	319.89	665.57	188.79	50.89	46.64	2.38
15	La Von reservoir	Texas	CE	29	1177.36	352.41	799.65	253.26	63.89	32.14	3.97
16	Hulah reservoir	Oklahoma	CE	13	1382.39	584.67	815.83	239.12	49.46	38.77	11.77

[a]Bureau of Reclamation (BR), US Department of the Interior Coast
[b]Soil Conservation Service (SCS), US Department of Agriculture
[c]Corps of Engineers (CE), US Department of the Army
[d]Forest Service (FS), US Department of Agriculture Geological
[e]Illinois state water survey

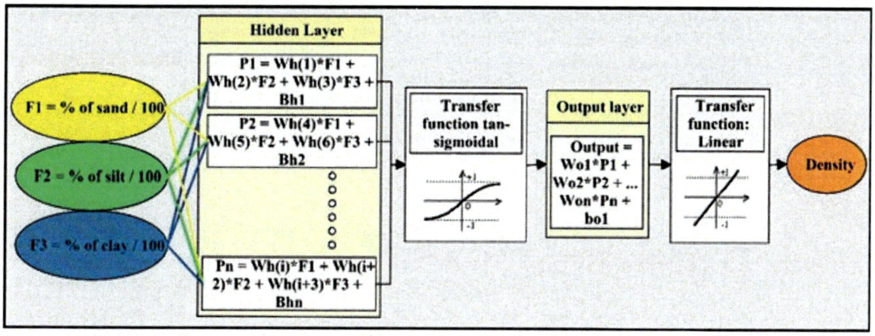

Fig. 17.1 Flowchart schematically presenting the neural network

functions (Jankowski and Duch 2001; Karlik and Olgac 2011; Yonaba et al. 2010). The transfer function of the hidden layer is selected as a tan-sigmoidal function (Eq. 17.3), and for the outputlayer, it is set as a linear function (Eq. 17.4).

$$sig(x) = \frac{1}{1 + e^{-x}} \tag{17.3}$$

$$lin(x) = mx \tag{17.4}$$

where x will be the output from the neurons, which is to be transferred, and m is the slope of the linear transfer function. The transferred value will be an input to the next layer or can be the output itself. The flowchart of the developed neural network is presented in Fig. 17.1.

17.3 Results and Discussion

For training, calibration, and testing process, data is to be divided into three partitions. In the present study, ANNs are developed using three sets of such partitioned data. Partitioning is carried by random selection of data. The first set has training data of 60%, 20% of calibration, and 20% of testing data (60, 20, and 20). The second set contains 70%, 15%, and 15% (70, 15, and 15) of training, calibration, and testing data. The third set comprises 80%, 10%, and 10% (80, 10, and 10) of training testing and calibration data. Number of neurons (n) is varied as three, four, five, and six for the hidden layer. The developed neural networks with varying data partition and number of neurons are presented in Table 17.4. The performance of the networks has been evaluated by comparing the predicted and observed densities. Nash–Sutcliffe model efficiency factor (E), Index of agreement (I), mean percentage error (P), and coefficient of determination (r^2) are the error statistics computed for all the stages (training, validation, and testing) and are presented in Table 17.5. Analyzing the data,

Table 17.4 Developed ANN details

Sr. No.	Number of data points			No. of neurons	Network Id
	Training	Calibration	Testing		
1	60	20	20	3	ANN1
2	60	20	20	4	ANN2
3	60	20	20	5	ANN3
4	60	20	20	6	ANN4
5	70	15	15	3	ANN5
6	70	15	15	4	ANN6
7	70	15	15	5	ANN7
8	70	15	15	6	ANN8
9	80	10	10	3	ANN9
10	80	10	10	4	ANN10
11	80	10	10	5	ANN11
12	80	10	10	6	ANN12

it is found that all the developed neural networks performed well; the best network based on the performance is ANN5. The regression plot of the observed and ANN predicted density is presented in Fig. 17.2.

Performance of the empirically derived density (Lara and Pemberton 1965) estimate and neural network predicted density estimate can be noted from the coefficient of determination. The coefficient of determination between Lara and Pemberton (1965) derived density and observed density is found as 0.6980, while the observed and ANN5 derived density has a coefficient of determination of 0.7285 (Fig. 17.3). The Nash-Suctliff model efficiency factor (E) and Index of agreement (I) obtained by comparing the observed and ANN5 model derived density are 0.91 and 0.73, respectively, while E and I obtained by comparing the observed and Lara and Pemberton (1963) model are 0.68 and 0.89 respectively. Thus, the developed ANN5 model is recommended to be used further to predict sediment density (for the reservoirs in which sediments always remained submerged). ANN model as a neural network code is given in Annexure 1. The code contains the weights and biases of the trained network, which is a ready to use MATLAB script.

17.4 Conclusion

In the present study, artificial neural network (ANN) models were examined to estimate reservoir deposited sediment density. For achieving this objective, 262 observed field densities and their constituent sand, silt, and clay proportions were considered. Comparison of the ANN model results with Lara and Pemberton (1963) model indicated that the ANN model (ANN5) is better than Lara and Pemberton (1963)

Table 17.5 Comparison of observed and predicted submerged reservoir sediment densities

Network ID	Nash Sutcliffe model efficiency, *E*			Index of agreement, *I*			Coefficient of determination, R^2			Mean percentage error, *P*		
	Training	Validation	Testing	Training	Validation	Testing	Training	Validation	Testing	Training	Validation	Testing
ANN1	0.704	0.815	0.672	0.90	0.94	0.90	0.730	0.645	0.744	6.1	5.6	10.6
ANN2	0.588	0.717	0.466	0.91	0.94	0.87	0.736	0.577	0.679	6.1	5.7	11.6
ANN3	0.600	0.783	0.597	0.90	0.95	0.90	0.733	0.658	0.756	7.7	7.3	11.7
ANN4	0.562	0.720	0.538	0.89	0.93	0.88	0.725	0.644	0.764	10.1	9.7	14.2
ANN5	0 732	0.775	0.668	0.91	0.93	0.90	0.734	0.666	0.760	5.6	5.2	10.2
ANN6	0.571	0.699	0.537	0.90	0.92	0.89	0.709	0.648	0.768	8.7	8.4	13.1
ANN7	0.613	0.650	0.572	0.92	0.92	0.90	0.732	0.646	0.779	5.3	5.0	8.2
ANN8	0.623	0.701	0.413	0.91	0.92	0.87	0.733	0.611	0.745	7.5	7.1	12.8
ANN9	0.631	0.626	0.642	0.88	0.88	0.90	0.634	0.627	0.713	5.6	4.8	11.2
ANN10	0.334	0.489	0.619	0.83	0.88	0.90	0.557	0.651	0.754	9.9	10.0	14.9
ANN11	0.140	0.459	0.385	0.80	0.86	0.82	0.492	0.633	0.647	12.7	11.0	20.3
ANN12	0.319	0.312	0.676	0.84	0.87	0.91	0.562	0.622	0.738	5.0	5.7	10.3

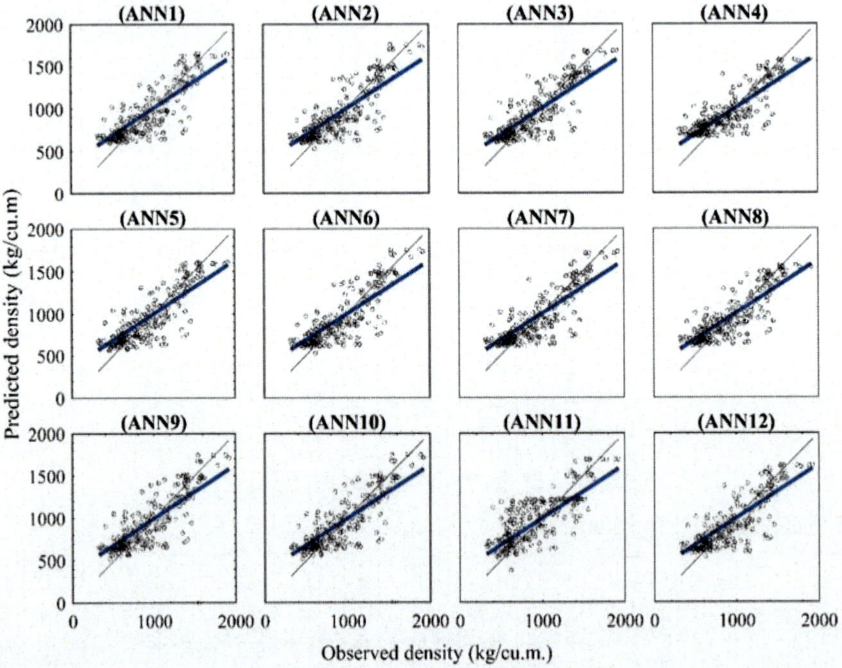

Fig. 17.2 Comparison of observed and ANN predicted densities (ANN1) $R^2 = 0.7222$, (ANN2) $R^2 = 0.7176$, (ANN3) $R^2 = 0.7279$, (ANN4) $R^2 = 0.7208$, (ANN5) $R^2 = 0.7285$, (ANN6) $R^2 = 0.7068$, (ANN7) $R^2 = 0.7284$, (ANN8) $R^2 = 0.7215$, (ANN9) $R^2 = 0.6515$, (ANN10) $R^2 = 0.7191$, (ANN11) $R^2 = 0.6166$, (ANN12) $R^2 = 0.7210$; the black dotted line is the best fit line and the blue line is the linear regression fit

model. The Nash–Sutcliffe model efficiency factor (E) and coefficient of determination (r^2) for Lara and Pemberton (1963) derived density estimates are 0.68 and 0.6980, respectively, which increased for ANN derived density estimates to 0.91 and 0.7285, respectively. Thus, the results confirm that the ANN5 model is flexible and robust to capture the complex sediment deposition process and is better than the Lara Pemberton empirical relationship.

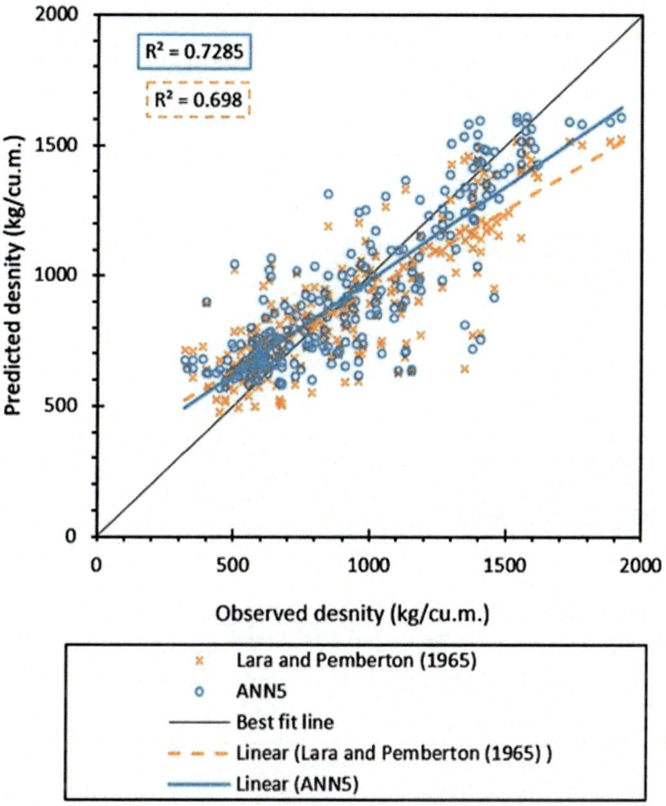

Fig. 17.3 Predictability of ANN5 derived density with the Lara Pemberton (1965) derived density

Annexure 1

```
function [y1] = DesnityPredictionFunction(x1)
%Density prediction function.
% [y1] = DensityPredictionFunction(x1) takes these arguments:
%   x = 3xQ matrix, input #1
% and returns:
%   y = 1xQ matrix, output #1
% where Q is the number of samples.
% ===== NEURAL NETWORK CONSTANTS =====
% Input 1
x1_step1.xoffset = [0;1;0];
x1_step1.gain = [0.0217391304347826;0.0233918128654971;0.0208333333333333];
x1_step1.ymin = -1;
% Layer 1
```

```
b1 = [-1.27683001570156018 37;-2.012427608471574203;-2.546392831139950 6299];
```

References

Annandale GW, Morris GL, Karki P (2016) Extending the life of reservoirs: sustainable sediment management for dams and run-of-river Hydropower. The World Bank. https://doi.org/10.1596/978-1-4648-0838-8

Basheer IA, Hajmeer M (2000) Artificial neural networks: fundamentals, computing, design, and application. J Microbiol Methods 43(1):3–31. https://doi.org/10.1016/S0167-7012(00)00201-3

Bowden GJ, Maier HR, Dandy GC (2002) Optimal division of data for neural network models in water resources applications. Water Resour Res 38(2):2–1–2–11. https://doi.org/10.1029/2001WR000266

Bussi G, Rodríguez-Lloveras X, Francés F, Benito G, Sánchez-Moya Y, Sopeña A (2013) Sediment yield model implementation based on check dam infill stratigraphy in a semiarid Mediterranean catchment. Hydrol Earth Syst Sci 17(8):3339–3354. https://doi.org/10.5194/hess-17-3339-2013

Donald F, Glauber R (1967) Multicollinearity in regression analysis: the problem revisited author(s). In: Farrar DE, Robert R. Glauber source : the review of economics and statistics, vol 49. no. 1. The MIT Press Stable UR, pp 92–107. https://www.jstor.org/stable/1937887

Jabbar, Yazad C, Yadav SM (2019b) Development of reservoir capacity loss model using bootstrapping of sediment rating curves. ISH J Hydraulic Eng 1–13. https://doi.org/10.1080/09715010.2019.1665483

Jabbar YC, Yadav SM (2019a) Development of a relationship between hydrometric and hydrographic observations to predict reservoir capacity loss. Acta Geophysica 67(5):1451–1469. https://doi.org/10.1007/s11600-019-00323-0

Jankowski N, Duch W (2001) Transfer functions: hidden possibilities for better neural networks. In: 9th European symposium on artificial neural networks, vol 7. pp 81–94

Karlik B, Olgac V (2011) Performance analysis of various activation functions in generalized MLP architectures of neural networks. Int J Artif Intell Expert Syst (IJAE) 1(4):111–122

Lane EW, Koelzer VA (1943) Density of sediments deposited in reservoirs. St Paul US Engineer District Sub-Office, Hydraulic Laboratory, University of Iowa, Iowa. https://water.usgs.gov/fisp/docs/Report_9.pdf

Lara, Pemberton EL (1963) Initial unit-weight of deposited sediments. In: Federal inter-agency sedimentation conference. U.S.D.A., USA, pp 818–845. file://catalog.hathitrust.org/Record/000212230%0A http://hdl.handle.net/2027/mdp.39015017878425

Lara JM Pemberton EL (1965) Initial unit weight of deposited sediments. In: Federal inter-agency sedimentation conference. U.S. Agriculture Research Service Publ 970, 818–845

Liu QJ, Shi ZH, Fang NF, Zhu HD, Ai L (2013) Modeling the daily suspended sediment concentration in a hyperconcentrated river on the Loess Plateau, China, using the Wavelet-ANN approach. Geomorphology 186:181–190. https://doi.org/10.1016/j.geomorph.2013.01.012

Maier HR, Dandy GC (2000) Neural networks for the prediction and forecasting of water resources variables: a review of modelling issues and applications. Environ Model Softw 15(1):101–124. https://doi.org/10.1016/S1364-8152(99)00007-9

Marineau MD, Wright SA (2017) Daily reservoir sedimentation model: case study from the Fena Valley reservoir, Guam. J Hydraul Eng 143(9):05017003. https://doi.org/10.1061/(ASCE)HY.1943-7900.0001344

Miller CR (1953) Determination of the unit weight of sediment for use in sediment volume computations. US Department of the Interior, Bureau of Reclamation, Project Planning Division

Noori N, Kalin L (2016) Coupling SWAT and ANN models for enhanced daily streamflow prediction. J Hydrol 533:141–151. https://doi.org/10.1016/j.jhydrol.2015.11.050

Raman H, Sunilkumar N (1995) Multivariate modelling of water resources time series using artificial neural networks. Hydrol Sci J 40(2):145–163. https://doi.org/10.1080/02626669509491401

Safari MJS, Aksoy H, Mohammadi M (2016) Artificial neural network and regression models for flow velocity at sediment incipient deposition. J Hydrol 541:1420–1429. https://doi.org/10.1016/j.jhydrol.2016.08.045

Strand RI, Pemberton EL (1982) Reservoir sedimentation technical guidelines for Bureau of reclamation. Denver, Colorado

Tokar AS, Johnson PA (1999) Rainfall-runoff modeling using artificial neural networks. J Hydrol Eng 4(3):232–239. https://doi.org/10.1061/(ASCE)1084-0699(1999)4:3(232)

Yang CT, Marsooli R, Aalami MT (2009) Evaluation of total load sediment transport formulas using ANN. Int J Sedim Res 24(3):274–286. https://doi.org/10.1016/S1001-6279(10)60003-0

Yegnanarayana B (2009) Artificial neural networks. PHI Learning Pvt. Ltd

Yonaba H, Anctil F, Fortin V (2010) Comparing sigmoid transfer functions for neural network multistep ahead streamflow forecasting. J Hydrol Eng 15(4):275–283. https://doi.org/10.1061/(ASCE)HE.1943-5584.0000188

Chapter 18
Micro-hydro Power Generation in India—A Review

Aparna M. Deulkar, Vivek S. Chavhan, and Pankaj R. Modak

Abstract Hydropower technology has been around for more than a century. Hydropower comes from converting the energy in flowing water—using a water wheel or a turbine—into useful mechanical power. This power is then converted into electricity by an electric generator. Micro-hydropower systems are small hydropower plants that have an installed power generation capacity of less than 100 kilowatts (KW). Many micro-hydropower systems operate "run of river," which means that no large dams or water storage reservoirs are built, and no land is flooded. Depletion of fossil fuel and the inability to meet the rising demand of electricity is some drawbacks for the economic development of India. This paper presents the study to investigate the possibility of the micro-hydro power generation and its advantages in India.

Keywords Micro-hydro power · Economic development · Installed power generation capacity · Renewable source

18.1 Introduction

Energy is very important for all human activity. Development of country is depending on available energy resources. Water is a natural source to generate the energy. In the past decades, there is a world-wide problem of fossil fuel depletion, climatic change, and increased electricity demand. (Anandh and Vinoth 2018) There is a fast depletion of renewable sources that was used in the past for the generation of electricity, and the difficulty in reachability of the grid supply to the remote villages was a big challenge

A. M. Deulkar (✉) · V. S. Chavhan · P. R. Modak
Department of Civil Engineering, AISSMS COE, Kennedy Road, Near R. T. O. Office, Pune 411001, India

V. S. Chavhan
e-mail: vschavhan@aissmscoe.com

P. R. Modak
e-mail: prmodak@aissmscoe.com

R. Jha et al. (eds.), *Water Resources Management and Reservoir Operation*,
Water Science and Technology Library 107,
https://doi.org/10.1007/978-3-030-79400-2_18

faced. The best possible remedial measure in this scenario is to make use of some renewable energy source like hydro power, wind, biomass, etc., so that it will be first step to reduce fossil fuel depletion. The aim of present work is to focus on micro-hydro power generation in India as renewable energy source to fulfill energy demand. Study of micro-hydro power plant focuses on three main folds such as technical as well as economical feasibility studies, design of civil works, and selection of electro mechanical components. There is a huge potential to develop a micro-hydro power plant which would meet the energy demand of the tribal settlement in India and thereby improving their living condition.

By considering advantages of micro-hydro power plant like no air pollution, no waste product like nuclear power plant, water left from the dam or cannel can be reuse, it provide provision for flood, and it is a clean source of energy. For rural development and employment, micro-hydro power plant is the best solution. Since small hydro power plants do not require long-term planning, less expensive, for rural electrification over a complete lifetime, they are cheapest technology available. It is noted every time that the development of large hydro power plant is resisted by local community, environmentalist, and NGOs due to issues such as deforestation and resettlement of the community (Höffken 2014). Another advantage of micro-hydro power plant is they use available flow of river, and no storage is required.

India is blessed with large number of river flowing over a country. All these rivers have great potential for hydro power which can be harnessed. Figure 18.1 shows the basin wise potential of Indian River system (Chauhan and Vig 2017).

India is ranked third biggest in the USA, with 5.5% global share in 2016 (Anandh and Vinoth 2018). Since India is largely dependent on fossil fuel imports to meet its energy demands, about 70% of India's electricity generation capacity is from fossil fuels (Anandh and Vinoth 2018). To meet the future energy demands, it is essential to tap all possible sources of small hydro energy using decentralized power generation (Anandh and Vinoth 2018). Micro-hydro power is the small-scale harnessing of energy from falling water which converts hydraulic energy to electric energy. It is cheapest solution for poor communities in rural areas with an affordable, easy

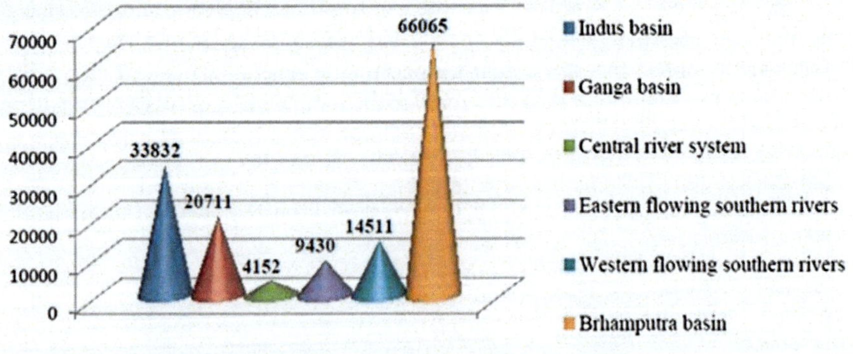

Fig. 18.1 Estimated hydro potential

to maintain, and long-term solution to their energy needs. In India, the total hydroelectrical potential is about 69% (including both large and small hydro projects and those installed/under installation). Anandh and Vinoth (2018) According to the international energy agency (IEA), large-scale hydroelectric plants currently supply 16% of the world's electricity. However, such kind of projects requires tremendous amounts of land impoundment, dams and flood control, and often they produce environmental impacts (Singh 2009). Among the renewable energy source, small hydro power contributes 13% of the total grid connected power generation, thereby constituting second largest grid connected system after wind power, as per the report by Ministry of New and Renewable Energy. Michael, and Jawahar (2017) Micro-hydro power is a type of hydroelectric power that typically produced up to 5–100 KW of electricity using the natural flow of water. This type of power plant can provide power to a small community.

Up to, November 2016, the installed capacity of hydropower in India is 43133.43 MW. This installed capacity of hydropower is about 13.97% of the total installed capacity of the country. The current potential of hydropower in India is about 1,48,701 MW. This potential is 3.45 times the installed capacity (Chauhan and Vig 2017). Indian power is generally based on fossil fuel to move toward renewable energy source and as 13.69% (Chauhan and Vig 2017) is the contribution by hydro power plant. To promote micro-hydro power plant by considering advantage above mentioned, it is necessary to take step toward micro-hydro power plant. Hence, this paper gives a review of micro-hydro power generation in India the water resources, current status, potential, and future of hydro energy in India.

18.2 Literature Review

This part is compiled with a review of past research work in the field of micro-hydro in India. Purpose of this literature review is to find key for further research in this field.

1. Anandh and Vinoth (2018) studied current scenario and future potential of small hydropower plant India. Share of hydro power in our country is declining persistently, from 46% in the year 1960 to 16.33% in the year 2015. Small-scale hydropower is a concentrated energy source with a long-lasting and robust technology, and its systems' life can extend to up to 50 years or more without any major new investments.
2. Dinesh Chauhan and Sunny Vig present paper gives a complete review of the water resources, current status, potential, and future of hydro energy in India. India ranks third after China and USA in the world in terms of numbers of dams. But large hydro power plant required many years for full construction and its commissioning of plant. Author suggest that besides the large hydro potential, small hydro power plant is the best solution to fulfill rural energy

demand by utilizing present hydro potential, and it also provide employment to rural population.
During the study, author observed that hydro generating unit sizes has been increased from 22 MW (from the independence) to 250 MW till today. Till august 2016, only 34.29% of the northern region of India potential and near about 2% of the north eastern region of India potential has been utilized. If the hydro power potential in these region is utilized completely, then nation's dependence on thermal power plants can be reduced significantly.

3. Adamantia et al. (2017) studied about climate change with respect to growing demand of energy. Globally averaged temperatures in 2016 were 0.99 °C (1.78 °F). By studying advantages of renewable energy, i.e., micro-hydro power plan authors provide a comprehensive assessment of the environmental impacts of small hydropower plants (SHPs). Considered the problem of assessing the environmental cost of a micro-hydro power (MHP) by using the approach of external cost. The external cost was derived from the combination of life cycle air pollutants emission factors for the three main construction components (concrete, steel, aggregates). The external cost was estimated to play a non-negligible part in the total investment cost that can affect the feasibility of the scheme. Also, more applications of the method on renewable energy sources will better demonstrate its benefits and drawbacks through additional practice.
4. Michael and Jawahar (2017) studied energy scenario. According to study, at the end of the Aug 2014, the power generation reached around 20,000 Mw only by small hydro power plant with more than 15,000 MW potential remaining unexploited. Authors consider Kerala state of India for study, which is having seasonal discharge but more discharge in rainy season. Tribal village in Kerala state does not have access to electricity. They use kerosene for lighting purpose. To fulfill energy demand of tribal village, first energy requirement and energy potential are calculated. Available discharge is reduces by 80% to account for various losses. By considering losses, electro mechanical equipment (like Turbine, Generator) is selected for power generation of 120 family. This study resulted with micro-hydro power plant is found to be technically and economically viable as operation cost is 60 Lakhs.
5. Manual (2015) gives brief idea, advantages, and challenges associated to the micro-hydropower. India is endowed with rich hydropower potential to the tune of 148 GW. India ranks fifth in the world in terms of usable hydro power potential. India has around 36 GW of installed hydropower capacity, whereas an additional 13 GW is under construction.
6. Saxena (2013) studied potential of small hydro power plant in India. According to his study potential of small hydro (upto 25 MW station capacity) in India is of about 20,000 MW of which about 3632 MW has been exploited. In India, hydro projects up to 25 MW station capacity have been categorized as SHP projects and responsibility of small hydro development rests with Ministry of New and Renewable Energy (MNRE). India has a history of about 110 years of hydropower. The first small hydro project of 130 KW commissioned in the hills of Darjeeling in 1897 marks the development of hydropower in India.

The total hydroelectric power potential in the country is assessed at about 150,000 MW, equivalent to 84,000 MW at 60% load factor. The potential of SHP projects is estimated at about 20,000 MW. Of this, 6474 potential sites with an aggregate capacity of 19,749 MW have been identified. Today the SHP program in India is essentially private investment driven. In fact, 329 private sector SHP projects of about 1748 MW capacity have been setup. Private sector entrepreneurs are finding attractive business opportunities in small hydro and state governments also feel that the private participation may be necessary for tapping the full potential of rivers and canals for power generation. Author concludes that appropriate selection of sites and sizing of projects to give higher plant load factors are considered important to further improve economic viability of commercial SHP projects.

7. Bhoi and Ali (2014) studied hydro power plant in India and its environment impact also studied HYDRO power plant projects in India. Hydro power can be classified as: large hydro power, medium hydro power, and small hydro power. These are classified according to the power generation capacity. Large hydro power: >100 MW, medium hydro power: 30–100 MW, small hydro power: 1–30 MW. Very small-scale hydro power plant classified as: Mini-hydro power whose capacity is between 100 KW and 1 MW and micro-hydro ranging up to 100 KW. Authors studied component part of hydro power plant along with turbine types according to head. Kaplan and Propeller $2 < H < 40$ Francis, $10 < H < 350$ Pelton, $50 < H < 1300$ (head in m). In environmental impact, author explains advantages and problem associate with hydropower like plant require suitable site, weather condition, local habitat, climatic condition, flow of water, and head. A large part of the land area is required to install a hydro project so may create disturbance to the local habitat. Requirement of large area is meeting by the afforestation programs which disturb the ecosystem. Sometime many aquatic animals are also affected by the construction of dam across water mass.
8. Saxena and Kumar (2010) studied hydropower development in India. In which they studied whole country is divided into five power regions, and planning is done on a regional concept. At the time of independence in the year 1947, only 1362 MW of electricity was produced in India. After that installed capacity of power generation had grown to 164,509 MW of which hydro is 37086 MW (25%), thermal is 106,433 MW (65%), nuclear is 4560 MW (2.9%), and renewable energy sources 16429 MW (7.7%). The share of small-scale hydropower (SHP) is 2820 MW. The potential of small hydro power projects is estimated at about 15,384 MW with 5718 potential identified sites.
9. Baidya (2006) studied Indian scenario of small hydro power and its advantages as a renewable resource. Author explains importance of small hydro power (SHP) and how it is optimally and viably source of energy to improve economic condition of the people in the village area and overall development of the country. This in turn help in capacity addition to the hydro power generation, and shortage of electricity can be avoided to some extent.

10. Vyas et al. (2015) studied micro-hydel power system design and its implementation in Rajasthan. In this study, authors observed that though India blessed with hydro potential, only marginal amount of power is so far tapped from the renewable source, because these potential sites are located in difficult terrains and remote access. Considering the design and the cost estimation, it can be concluded that the small hydro projects are feasible in the arid and semi-arid regions of Rajasthan. Author suggested that by considering important aspects like proper maintenance, providing silt excluders we can improved micro-hydro potential as renewable source.
11. Choudhury and Ghosh (2013) focus on responsible hydro power development in India. As hydro power is important and renewable source of energy, but this article is focus on "responsible" hydropower development. Means hydropower development should more stable and sustainable investment for medium-to-long term.
12. Souza and Donald (2015) studied need for hydropower in India, hydro potential, and growth and share of hydroelectric installed capacity and generation. Share of hydro power in India is 16.9% of the total installed capacity of 237,742.94 MW as on February 28, 2015, and 4.4% global install capacity and ranked 6th in the list of global nations. The hydropower generation for 2012–12 and 2013–14 stood at 12.5 and 14% of the total energy generation. As against the power generation target of 122,263 MU for 2013–14, generation from hydroelectric power stations (above 25 MW Installed Capacity) was 134,847.52 MU, which was 10.29% more than the target.

18.3 Conclusion

During literature study, it was observed that micro-hydropower is the best solution to overcome growing energy demand. Some work has been carried out related to the same, but there is need to more focus on micro-hydro power development in sustainable way in state of dependency on hydropower. It is observed that to improve livelihood of villagers and also remote villages' micro-hydro power is the solution.

Hydro is a renewable resource that is replenished by the environment over a relatively short period of time. Water is neither depleted nor its composition altered during the generation cycle. A run of river plant stores water in the weir for a short duration, and the water is returned to water cycle on each day. Small hydro projects (90%) efficient in utilization of the resource than solar (15–20%), wind (35%) and other renewable energy sources. In long term, small hydro schemes have the least impact among the environmental indicators like acidification, climate change, ozone layer depletion, photo chemical oxidation (smog), etc. Capital subsidy for small hydro power should be increased, and better mechanism to deliver the subsidy should be devised. Incentive available to small hydro projects up to 25 MW may be extended to the hydro power projects up to 100 MW. Small hydro power plants have advantage of life span almost 50 years, i.e., more than twice the life span of other renewable sources like wind, solar, etc.

References

Adamantia ZV, Didaskalou E, Georgakellos D (2017) Article on financial appraisal of small hydropower considering the cradle-to-grave environmental cost: a case from Greece
Anandh T, Vinoth R (2018) A comprehensive assessment of small hydro power in India—current scenario and future potential. J Mech Prod Engg 8: 413–424
Baidya G (2006) Development of small hydro. Himalayan small hydropower summit. pp 34–43
Bhoi R, Ali SM (2014) Potential of hydro power plant in India and its impact on environment. J Eng 10:114
Chauhan D, Vig S (2017) A review of present status and potential of hydro power In India. Int J Mod Trends Eng 4:2349–9745
Choudhury N, Ghosh A (2013) Responsible hydropower development in India. Manual by the Council on Energy, Environment and Water for the Independent Power (EEW)
Höffken JI (2014) A closer look at small hydropower projects in India: Social acceptability of two storage-based projects in Karnataka. pp 155–166
Hydro power in India key enables for a better tomorrow (2015) Pwc manual. www.pwc.in
Maual Saxena P (2013) Renewable energy akshayUrja 6:24
Michael PA, Jawahar CP (2017) Design of 15 KW micro hydro power plant for rural electrification at valara. In: 1st International conference on power engineering, computing and control, PECCON, pp 163–171
Saxena P, Kumar A (2010) Hydropower development in India. In: Conference alternate hydro energy centre, Indian Institute of Technology Roorkee
Singh D (2009) Micro-hydro-power, resource assessment handbook, an initiative of the Asian and Pacific Center for Transfer of Technology
Souza SMD, Donald J (2015) Green growth and hydropower in India. Draft by The Energy and Resources Institute (Teri)
Vyasa A, Guptab NK, Guptac SK, Gautamd P, Jehooe AS (2015) Mini/micro hydel power system design and its implementation in Rajasthan. In: International conference on water resources, coastal and ocean engineering (Icwrcoe), Aquatic Procedia 4:1537–1544

Chapter 19
Runoff Simulation and Irrigation Water Requirement for Barman Command

A. Vishwakarma, M. K. Choudhary, and M. S. Chauhan

Abstract The study is based on runoff simulation using the SIMHYD and AWBM model of the Narmada River at Barman. The NSE and R^2 values of AWBM for calibration are 0.751 and 0.821 and for the validation are 0.797 and 0.862, respectively, while the NSE and R^2 values of SIMHYD for calibration are 0.814 and 0.731, and for validation period 0.700 and 0.755, respectively. AWBM accuracy values indicate a better agreement between the observed and simulated runoff than the SIMHYD model. It is concluded that AWBM is more suitable for Barman command to simulate the basin's hydrological response to the rainfall and predict daily runoff with a better degree of accuracy. Further study is carried out to estimate the irrigation water requirement based on rainfall pattern knowledge to achieve the proper crop calendar of Barman command. For estimating the irrigation water requirement of a wheat crop, 12 years (2001–2012) average climatic data, including 12 years (2001–2012) of annual rainfall data, is used. The probability analysis of rainfall was (at 20, 50, and 80%) conducted to better understand rainfall behavior. CROPWAT model is used to calculate the reference evapotranspiration (ET_O) by using climatic parameters (like maximum and minimum temperature, relative humidity, wind velocity, and sunshine hours). It is found that the rainfall for dry, wet, and at normal probability year are 598 mm, 1168 mm, and 866 mm, respectively, and corresponding irrigation water requirement estimated by CROPWAT are 274 mm, 264.7 mm, and 269.5 mm.

Keywords American water balance model (AWBM) · Coefficient of determination (R^2) · Nash–Sutcliffe efficiency (NSE) · Reference evapotranspiration · Irrigation water requirement · CROPWAT 8.0

A. Vishwakarma (✉) · M. K. Choudhary · M. S. Chauhan
Department of Civil Engineering, Maulana Azad National Institute of Technology, Bhopal, India

R. Jha et al. (eds.), *Water Resources Management and Reservoir Operation*,
Water Science and Technology Library 107,
https://doi.org/10.1007/978-3-030-79400-2_19

19.1 Introduction

Estimation of direct runoff is a complicated and challenging process. Nowadays, a large number of rainfall-runoff models are available with different levels of complexity. A rainfall–runoff model is a mathematical model that describes catchment and gives the relationship between precipitation and runoff (Podger 2004). However, models with simple structure and least input providing adequate results are most liked by the hydrologists (Boughton 2006). Many studies had been performed to evaluate the applicability of various rainfall–runoff models (Jones et al. 2006; Thimme et al. 2013; Haque et al. 2015; Li et al. 2015; Yu 2015; Sachan et al. 2016; Chouhan et al. 2016). In this study, two conceptual models, namely AWBM (Boughton 2004; Boughton and Chiew 2007) and SIMHYD (Chiew and McMahon 1991; Chiew and Siriwardena 2005) are applied for the runoff simulation of the Barman command of Madhya Pradesh, India.

Rainfall is not a constant phenomenon. It occurs as normal, deficit, and heavy as its magnitude may be daily, monthly, or yearly (Manivannan et al. 2016). Hence, for a complete analysis of rainfall, statistical approach is preceded for the Barman command of Madhya Pradesh. CROPWAT model (Smith 1992) is used to calculate the irrigation water requirement of the rabi crop of the study area for variable rainfall conditions. Hence, it is concluded that how much water should be released from the proposed dam upstream of the study area as per the rainfall condition of a particular year.

19.2 Study Area and Data

19.2.1 Study Area

The Barman is a town on the Narmada River banks located at a longitude of 79°01′19″ E and latitude of 23°01′33″ N in the Narsinghpur district, Madhya Pradesh, India, as shown in Fig. 19.1. In the present study, runoff simulation and irrigation water requirements of the Basin of Narmada River of Barman command have been assessed.

19.2.2 Data Collection

Daily rainfall data of 12 years from 2001 to 2012 has been collected. Reference evapotranspiration (ETo) has been calculated using climatic data of Barman for the period from years 2001–2012 by the CROPWAT model. It was assumed that the average climatic data of Barman station represents its whole area. The ETo was low in December (2.21 mm/day) and maximum (6.76 mm/day) during May, and the average ETo was found to be 3.84 mm/day. The average climatic data is shown in Fig. 19.2.

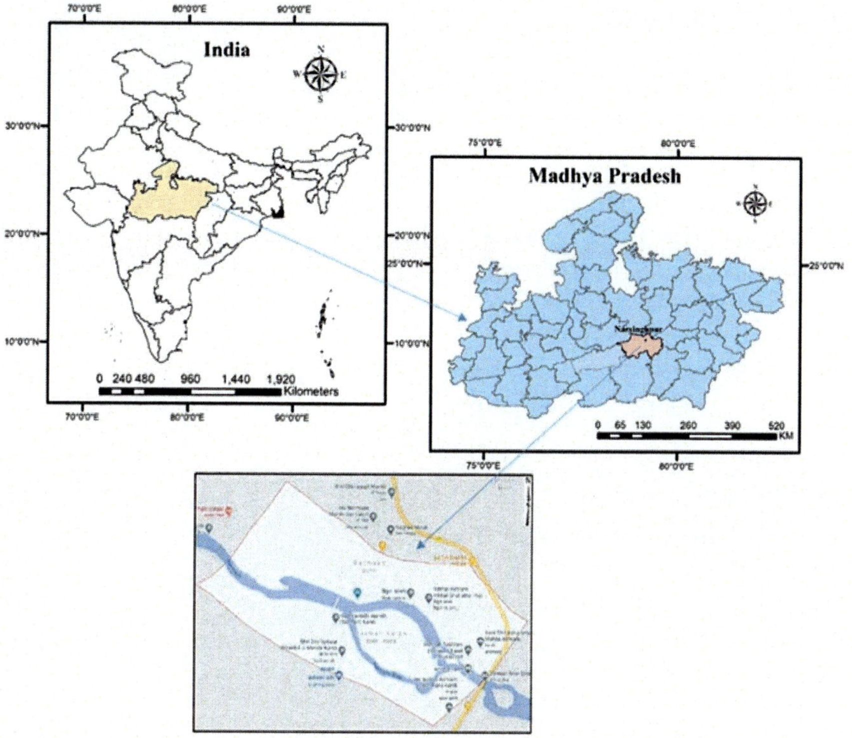

Fig. 19.1 Location of Barman in Madhya Pradesh, India

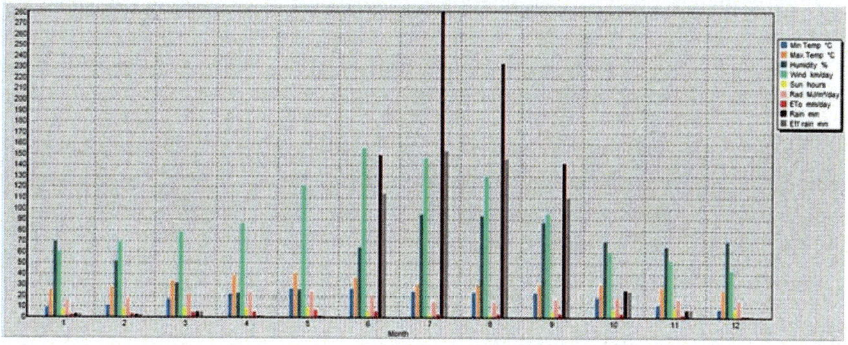

Fig. 19.2 Average climatic data of the study area

19.3 Methodology

19.3.1 Net Irrigation Water Requirement

As per FAO manual no. 56 (Allen et al. 1998), "The irrigation water requirement represents the difference between the crop water requirement and effective precipitation. The irrigation water requirement also includes additional water for leaching of salts and compensating for non-uniformity of water application." It can be summarized as,

$$IR_n = \sum_{t=0}^{T} (K_c \cdot ET_o - P_{\text{eff}}) \tag{19.1}$$

where:

IR_n Net irrigation requirement (mm).
ET_o Reference crop evapotranspiration (mm).
P_{eff} Effective dependable rainfall (mm).
K_c Crop coefficient.
T Total growing period of crops.

19.3.2 Rainfall–Runoff Modeling

Rainfall–runoff models give the relationship between precipitation and overland flow. Here, two models, AWBM and SIMHYD, were used to evaluate runoff simulation of Barman command in Narmada. The essential data required for the setup of the AWBM and SIMHYD model are listed below:

1. Area of the catchment in km^2
2. Rainfall data daily time series, mm/day
3. Potential evapotranspiration data daily time series, mm/day
4. Observed discharge data daily time series, m^3/s.

19.3.3 Probability Analysis

For proper planning and management of irrigation water, rainfall of normal, dry, and wet years can be calculated by using the statistical probability approach. The steps involved are:

1. Calculation of yearly rainfall as per the available rainfall records.
2. Arrangement of yearly rainfall in descending order.

3. Calculation of the plotting position or can say the probability of exceedance of rainfall by using Weibull's formula of statistical probability, given as:

$$Fa = m/(N + 1) \tag{19.2}$$

where

N number of records,
m rank number,
Fa plotting position.

4. Preparation of a graph on a semi-log scale between rainfall (mm/year) and its probability of exceedance.
5. Calculation of yearly rainfall values at 20, 50, and 80% probability.
6. Calculation of the monthly values of the normal year as per the following relationship:

$$P_{i\text{norm}} = P_{i\text{av}} * \frac{P\text{norm}}{P\text{av}} \tag{19.3}$$

where

$P_{i\text{av}}$ Average monthly rainfall of the ith month.
$P_{i\text{norm}}$ Monthly rainfall for the normal year of the ith month.
P_{av} Average yearly rainfall.
P_{norm} Yearly rainfall at 50% probability of exceedance.

Similarly, values for dry and wet years can be calculated.

19.3.4 Performance Evaluation of Model

19.3.4.1 Coefficient of Determination, R^2

Coefficient of determination is a method to evaluate the reliability of the model between the simulated and observed runoff data. Mathematically, it is expressed as follows:

$$R^2 = \left(\frac{\sum_{i=1}^{n} (O_i - \bar{O})(P_i - \bar{P})}{\sqrt{\sum_{i=1}^{n} (O_i - \bar{O})^2} \sqrt{\sum_{i=1}^{n} (P_i - \bar{P})^2}} \right)^2 \tag{19.4}$$

where O and P are observed and predicted values, respectively.

19.3.4.2 Nash–Sutcliffe Criteria

It is defined as “one minus the sum of the absolute squared differences between the calculated and observed values normalized by the variance of the observed values” during the study (Nash and Sutcliffe 1970).

Mathematically, the formula is expressed as:

$$\mathrm{NSE} = 1 - \frac{\sum_1^n (O_i - P_i)^2}{\sum_1^n (O_i - \bar{O})^2} \tag{19.5}$$

where

O_i is the observed discharge,
P_i is the modeled or predicted discharge,
$\bar{O}$ is the mean of the observed discharge.

19.4 Results and Analysis

19.4.1 AWBM Calibration and Validation Charts

The AWBM model was calibrated for five years period from 2001 to 2005 and then validated for the remaining period of three years from 2006 to 2008. The NSE graph presenting a comparison between observed and simulated discharge during model calibration (NSE = 0.751) and validation (NSE = 0.821) is shown in Fig. 19.3. The R^2 value obtained during calibration and validation is shown in Fig. 19.4 and Fig. 19.5, respectively.

19.4.2 SIMHYD Model

19.4.2.1 SIMHYD Model Calibration and Validation Charts

The SIMHYD model was calibrated for five years period from 2001 to 2005 and then validated for the remaining period of three years from 2006 to 2008. The NSE graph presenting a comparison between observed and simulated discharge during model calibration (NSE = 0.814) and validation (NSE = 0.731) is shown in Fig. 19.6. The R^2 value obtained during calibration and validation is shown in Fig. 19.7 and Fig. 19.8, respectively.

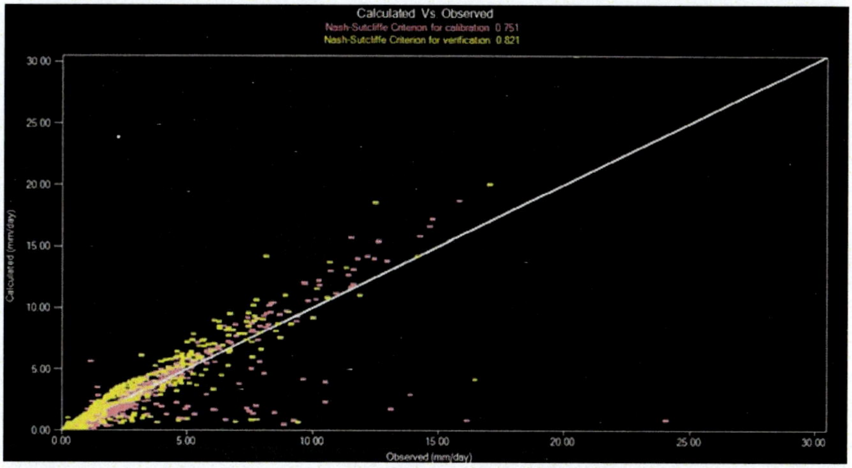

Fig. 19.3 Nash–Sutcliffe efficiency of AWBM calibration and validation

Fig. 19.4 R^2 chart for calibration

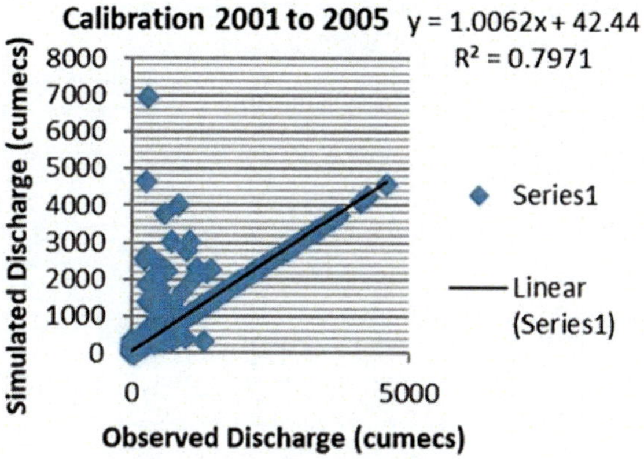

Fig. 19.5 R^2 chart for validation

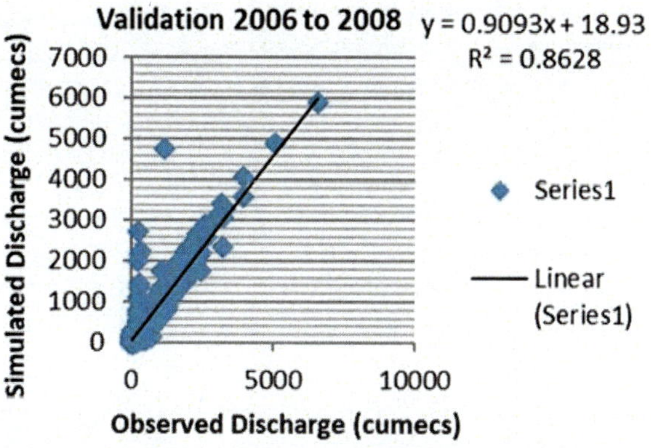

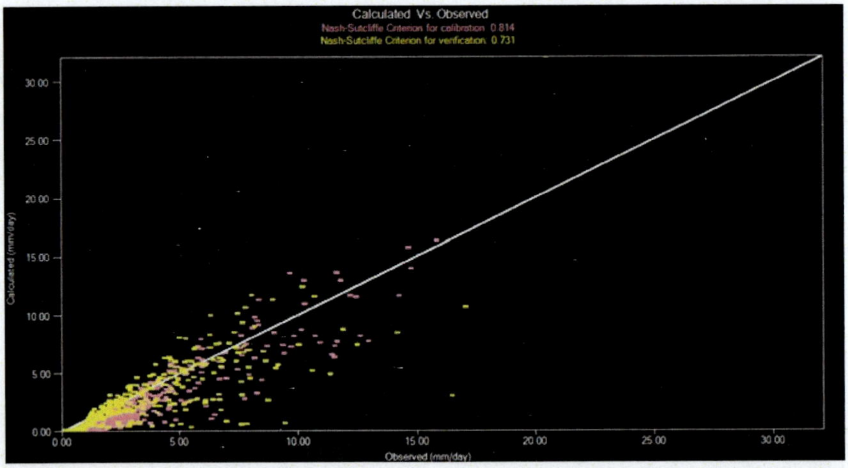

Fig. 19.6 Nash–Sutcliffe efficiency of SIMHYD model calibration and validation

Fig. 19.7 R^2 chart for calibration

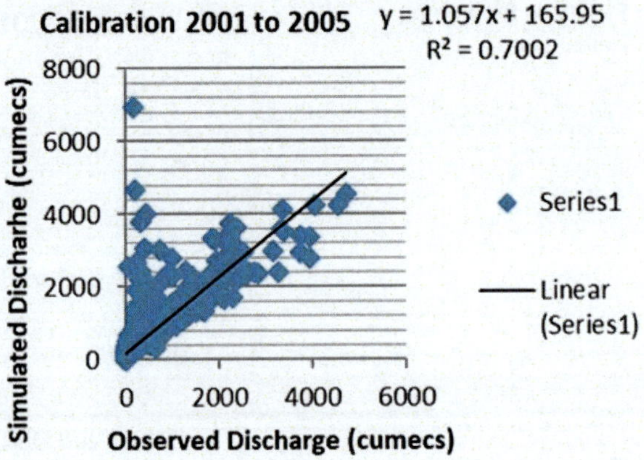

Fig. 19.8 R^2 chart for validation

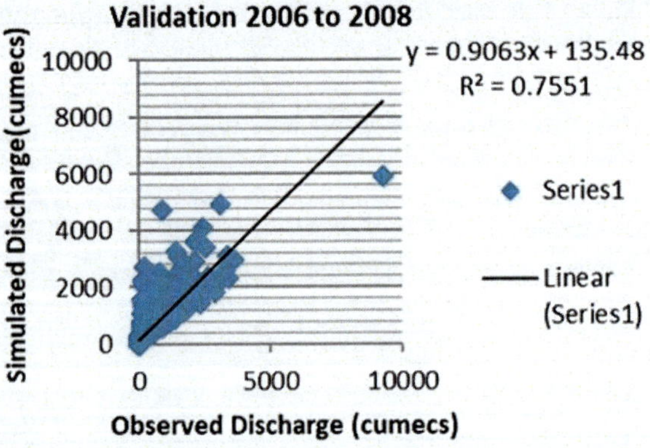

19.4.3 Accuracy of Models

The accuracy of both the models (AWBM and SIMHYD) for calibration and validation has been summarized in Table 19.1. AWBM performed better than the SIMHYD model.

19.4.4 Effective Rainfall

There are different methods exist to estimate the effective rainfall in the CROPWAT 8.0 model. USDA Soil Conservation Service Method is one of the most commonly used methods. The average monthly effective rainfall of the Barman is shown in Fig. 19.9.

19.4.5 Probability Analysis of Yearly Rainfall

Using Weibull's formula, 12 years of rainfall records of Barman arranged in descending order as shown in Table 19.2.

Table 19.1 Accuracy parameter values for calibration and validation

Model	Accuracy criteria	Calibration	Validation
AWBM	Nash–Sutcliffe efficiency (NSE)	0.751	0.821
AWBM	Coefficient of determination (R^2)	0.797	0.862
SIMHYD	Nash–Sutcliffe efficiency (NSE)	0.814	0.731
SIMHYD	Coefficient of determination (R^2)	0.700	0.755

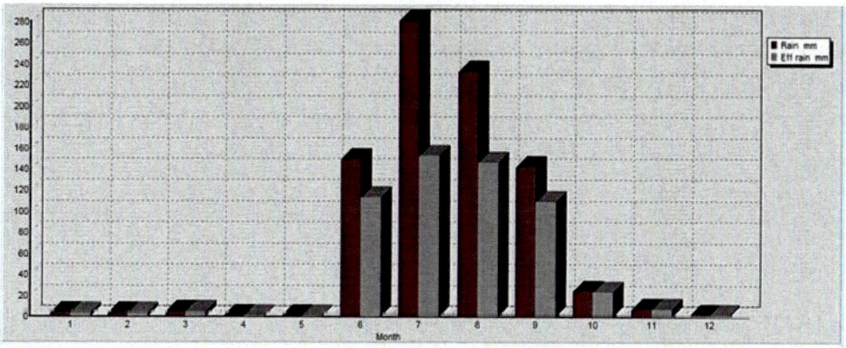

Fig. 19.9 Average monthly effective rainfall distribution of study area

Table 19.2 Rainfall records

Year	2012	2011	2008	2003	2009	2010	2007	2002	2005	2006	2001	2004
Rank no.	1	2	3	4	5	6	7	8	9	10	11	12
Rain (mm/year)	1267.6	1156.7	1033.6	993.8	949.1	936.2	931.3	836.1	725.7	500.0	460.5	444.5
Fa (%)	7	14	21	29	36	43	50	57	64	71	79	86

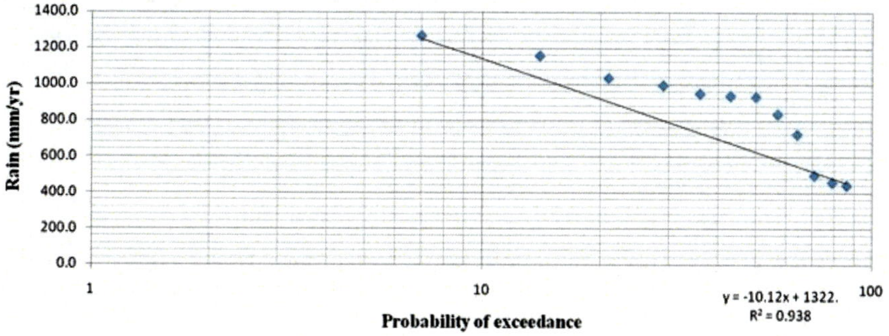

Fig. 19.10 Probability yearly rainfall

A logarithmic regression equation is obtained by preparing a semi-log graph between yearly rainfall and probability, which gives yearly rainfall values at 20, 50, and 80% probability of exceedance as 1168 mm, 866 mm, and 598 mm, respectively. The graph is shown in Fig. 19.10.

19.4.6 Probability Analysis of Monthly Rainfall

It is clear from the probability analysis that the maximum deficiency in rainfall in July is 95.1 mm, and the minimum deficit of 0.017 mm rainfall has been observed in April month as shown in Fig. 19.11.

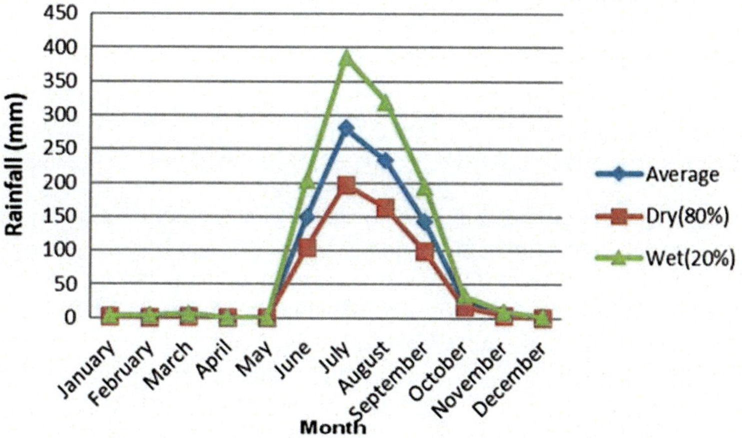

Fig. 19.11 Monthly rainfall under different probability level for Barman command

ETo station Barman | Crop Wheat

Rain station BARMAN | Planting date 15/11

Month	Decade	Stage	Kc	ETc	ETc	Eff rain	Irr. Req.
			coeff	mm/day	mm/dec	mm/dec	mm/dec
Nov	2	Init	0.30	0.86	5.1	1.1	4.2
Nov	3	Init	0.30	0.79	7.9	1.4	6.5
Dec	1	Init	0.30	0.73	7.3	0.9	6.4
Dec	2	Deve	0.36	0.79	7.9	0.2	7.7
Dec	3	Deve	0.64	1.44	15.8	0.5	15.3
Jan	1	Deve	0.93	2.14	21.4	0.9	20.5
Jan	2	Mid	1.13	2.65	26.5	1.0	25.5
Jan	3	Mid	1.14	2.97	32.7	1.0	31.7
Feb	1	Mid	1.14	3.27	32.7	0.9	31.7
Feb	2	Mid	1.14	3.57	35.7	0.9	34.7
Feb	3	Late	1.07	3.75	30.0	1.1	28.9
Mar	1	Late	0.82	3.20	32.0	1.6	30.4
Mar	2	Late	0.54	2.32	23.2	1.9	21.3
Mar	3	Late	0.34	1.59	6.4	0.5	5.7
					284.6	13.9	270.7

Fig. 19.12 IWR of wheat crop in Barman command

19.4.7 Irrigation Water Requirement

The irrigation water requirements in the Barman area have been computed using CROPWAT 8.0 software. The model calculates the irrigation water requirement of rabi crop (Wheat) on a ten-daily basis, as shown in Fig. 19.12.

As per the average annual rainfall data and average climatic data, the irrigation water requirement is 270.7 mm. Similarly, it can be calculated for rainfall at normal (50%), dry (80%), and wet (20%) year of probability.

19.5 Conclusions

The models were tested and evaluated based on accuracy criteria Nash–Sutcliffe efficiency and coefficient of determination (R^2). The NSE and R^2 values from AWBM for calibration are 0.751 and 0.821 and for the validation is 0.797 and 0.862, respectively, while NSE and R^2 values from SIMHYD for calibration are 0.814 and 0.731, and for validation are 0.700 and 0.755, respectively. AWBM accuracy values indicate a better agreement between the observed and simulated runoff than the SIMHYD model. Based on the accuracy values, it is concluded that AWBM is more suitable for the Barman command in simulating the hydrological response of the command to the rainfall and predicting daily runoff with the greater degree (13–14% higher for R^2 value) of accuracy.

It is concluded that the maximum contribution of rainfall is in July (around 30.8%) and then in August (about 28%). Maximum deficiency in monthly rainfall from average to dry year of probability is observed in July and minimum in April. Around 27% of the yearly rainfall deficiency is observed between average rainfall and rainfall in a dry year. It is found that the rainfall for dry, wet, and at normal probability year are 598 mm, 1168 mm, and 866 mm, respectively, and corresponding irrigation water requirement estimated by CROPWAT are 274 mm, 264.7 mm, and 269.5 mm, respectively. This study provides the idea related to irrigation water requirements at variable rainfall pattern satisfactorily. Hence, it becomes very flexible and easy to calculate rainfall at different probability levels by using statistical probability analysis and irrigation water requirements by using CROPWAT 8.0 model to achieve the proper irrigation planning.

Referencess

Allen RG, Pereira LS, Raes D, Smith M (1998) Crop evapotranspiration—guidelines for computing crop water requirements-FAO Irrigation and drainage paper 56. FAO Rome 300(9):D05109

Boughton W (2004) The Australian water balance model. Environ Model Softw 19(10):943–956

Boughton W (2006) Calibrations of a daily rainfall-runoff model with poor quality data. Environ Model Softw 21(8):1114–1128

Boughton W, Chiew F (2007) Estimating runoff in ungauged catchments from rainfall, PET and the AWBM model. Environ Model Softw 22(4):476–487

Chiew FH, McMahon TA (1991) The Applicability of Morton's and Penman's evapotranspiration estimates in rainfall-runoff modeling 1. JAWRA J Am Water Resources Association 27(4):611–620

Chiew FHS, Siriwardena L (2005) Estimation of SIMHYD parameter values for application in ungauged catchments 1

Chouhan D, Tiwari HL, Galkate RV (2016) Rainfall runoff simulation of Shipra river basin using AWBM RRL toolkit. Int J Civil Eng Tech Res (IJETR) 5(3):73–76

Haque MM, Rahman A, Hagare D, Kibria G (2015) Parameter uncertainty of the AWBM model when applied to an ungauged catchment. Hydrol Process 29(6):1493–1504

Jones RN, Chiew FH, Boughton WC, Zhang L (2006) Estimating the sensitivity of mean annual runoff to climate change using selected hydrological models. Adv Water Resour 29(10):1419–1429

Li L, Lambert MF, Maier HR, Partington D, Simmons CT (2015) Assessment of the internal dynamics of the Australian water balance model under different calibration regimes. Environ Model Softw 66:57–68

Manivannan S, Khola OPS, Dinesh D (2016) Probability analysis of weekly rainfall for crop planning in Nilgiris hills of Tamil Nadu. J Agromet 18(1):163–164

Nash JE, Sutcliffe JV (1970) River flow forecasting through conceptual models part I—a discussion of principles. J Hydrol 10(3):282–290. https://doi.org/10.1016/0022-1694(70)90255-6

Podger G (2004) Rainfall runoff library user manual. CRC for Catchment Hydrology, Melbourne

Sachan S, Chandola VK, Lohani AK (2016) Probability analysis of rainfall and crop water requirement using CROPWAT model for crop planning in a canal command of upper Bhima Basin of Maharashtra. Int J Agric Environ Biotechnol 9(1):123

Smith M (1992) CROPWAT: a computer program for irrigation planning and management (No. 46). Food & Agriculture Organization

Thimme Gowda P, Manjunaththa SB, Yogesh TC, Satyareddi SA (2013) Study on water requirement of Maize (Zea mays L.) using CROPWAT model in northern transitional zone of Karnataka. J Environ Sci Comput Sci Eng Technol 2(1):105–113

Yu B (2015) How would peak rainfall intensity affect runoff predictions using conceptual water balance models? Proc Int Assoc Hydrol Sci 371:109–115

Chapter 20
Nonlinear Regression Analysis Between Discharge and Head for Piano Key Weirs with Increasing Developed Length (L/W) Ratio and Constant Channel Width

Amiya Abhash and K. K. Pandey

Abstract Piano key weir (PKW) is becoming increasingly popular for dissipation of excess discharge from spillways in old and newly constructed hydraulic structures. The relationship between discharge and head thus warrants necessary attention. The relationship between discharge and head for piano key weirs with increasing developed ratio for free flow over PKW for a constant channel width is investigated. Further, the variation of increasing developed length ratio in constant channel width with the power exponent of the head is analyzed. Nonlinear regression analysis was performed using the Levenberg–Marquardt algorithm. The parameter estimate and ANOVA table were prepared for flow over piano key weirs from data available in the literature. The analysis shows a decreasing dependence on the exponent of the head with increasing developed length ratio for the same channel width.

Keywords Piano key weir · Free flow · Discharge · Head · Developed length ratio

20.1 Introduction

With increasing floods and to avoid the danger of overtopping, many spillways require replacement or an increase in their discharge capacity. This led to research for increasing the crest length of the weir to improve their discharge capacity. Piano key weir serves as an excellent alternative for increasing the overflow capacity of existing and new dams (Leite Ribeiro et al. 2009, 2011).

The piano key weir is an improvement over rectangular Labyrinth type weir, which has cantilevered apexes so that the allowable weir developed length can be fit in the previous spillway channel width (Anderson and Tullis 2012). This enables improved hydraulic capacity while reducing the structural footprint (Lempérière and Ouamane 2003; Lempérière and Vigny 2011). Piano key weir's geometric feature is

A. Abhash (✉) · K. K. Pandey
Department of Civil Engineering, Indian Institute of Technology (BHU), Varanasi, India
e-mail: amiyaa.rs.civ16@itbhu.ac.in

R. Jha et al. (eds.), *Water Resources Management and Reservoir Operation*,
Water Science and Technology Library 107,
https://doi.org/10.1007/978-3-030-79400-2_20

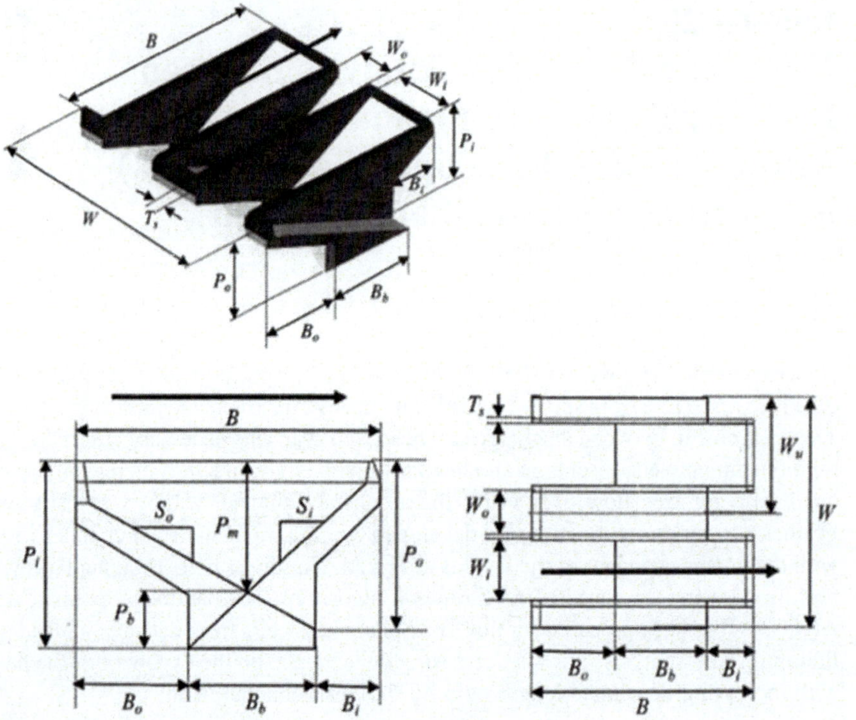

Fig. 20.1 Fundamental parameters of a PKW Pralong et al. (2011)

easily adaptable than a Labyrinth weir for direct placement on the dam crest (Pralong et al. 2011).

A naming convention was formulated for piano key weirs by Pralong et al. (2011). The basic unit of PKW consists of two half inlets and one outlet, as shown in Fig. 20.1.

As shown in Fig. 20.1, the main geometric parameters of the PKW consist of the total crest length L of weir, weir height P, channel width W, the transverse weir crest length B, the total upstream and downstream cantilever (overhang), lengths B_{o} and B_{i}, and the upstream and downstream widths W_{o} and W_{i}. Important fluid parameters include density ρ, dynamic viscosity μ, and surface tension σ.

20.2 Literature Review

The piano key weirs are hydraulically more efficient than linear overflow weirs or classical Labyrinth weirs for similar head and spillway width for flow in low heads (Laugier et al. 2012; Tiwari and Sharma 2015).

There is no generally accepted standardized design procedure of PKW currently available. This is due, in part, to a large number of PKW geometric parameters, and a limited understanding of their influences on the head-discharge relationship

(Anderson and Tullis 2011). The discharge from PKW is affected by more than 20 geometrical parameters (Sharma and Tiwari 2013; Tiwari and Sharma 2017), which are interlinked among each other.

Many researchers have given the discharge capacity of piano key weirs for unit width of channel and for a given head. They expressed the discharge head relationship as:

$$Q = KH^n,$$

where Q is discharge, H is head over the crest or the head at some distance upstream from piano key weir, n is the exponent, L is the developed length, and K is the system constant.

Lempériѐre and Ouamane (2003)obtained a head discharge relationship for a "Standard model" of PKW with all the characteristics defined by the only parameter Pm (Fig. 20.1). They also proposed a formula relating the specific discharge q with upstream total upstream head Hu. The value of exponent n was found to be unity.

Kabiri-Samani and Javaheri (2012) proposed dependence of the discharge over a PKW in case of free flow on the following factors

$$Q = f(H, L, P, W, W_i, W_o, B, B_i, B_o, g, \rho, \sigma, \mu) \tag{20.1}$$

where f is a function. Performing dimensional analysis, Eq. 20.1 was written as

$$C_d = \phi\ (H/P,\ L/W,\ B/P,\ W_i/W_o,\ B_i/B,\ B_o/B,\ W,\ R)\ldots \tag{20.2}$$

where ϕ is another functional symbol,

$$W = (\rho\ H/\sigma)\ 0.5Q/[(H + P)\ W]\ \text{and}\ R = \rho\ Q\ H/[(H + P)\ W\ \mu]\ \ldots \tag{20.3}$$

where R and W are the Reynolds and Weber numbers, respectively.

$H > 30$ mm was ensured so that the surface tension effect on discharge values is small. Since the flow is turbulent around PKW, the viscosity effect is considerably smaller compared with the gravity effect and hence discharge dependence on the Weber and Reynolds number can be omitted.

The discharge capacity of a PKW is given by the same formula as that of a sharp-crested weir as below:-

$$Q = \frac{2}{3}C_d\left(\sqrt{2g}\right)WH^{1.5}\ldots \tag{20.4}$$

where C_d is the coefficient of discharge, g is the acceleration due to gravity, and W is the channel width.

The exponent "n" in the formula above is taken as 1.5 subject to the limitations $H > 30$ mm, $0.1 \leq H/P \leq 0.6$, $2.5 \leq L/W \leq 7$, $1 \leq B/P \leq 2.5$, $0.33 \leq W_i/W_o \leq$

1.22, $0 \leq B_i/B \leq 0.26$, $0 \leq B_o/B \leq 0.26$, and $H_d/H \leq 0.6$. The data points from the discharge versus head relationship for different values of developed length ratio *L/W* were extracted from literature with plot digitizer. Exponent n was observed to be 1.5 in various literature (Anderson 2011; Laugier 2011; Leite Ribeiro et al. 2011; Kabiri-Samani and Javaheri 2012; Mehboudi et al. 2016). Head over exponent was interestingly observed to be 0.75 in the experiments done by Tiwari and Sharma (2017) with a developed length ratio (*L/W*) equal to 4.76. Discharge head relationship was observed as below:

$$\frac{Q}{LW} = CH^{3/4}\ldots \tag{20.5}$$

The head-to-weir height ratio (*H/P*) was kept below 0.1 for the experiments while greater than 0.1 for experiments done by Kabiri-Samani and Javaheri (2012). However, the discharge intensity was observed to be greater in this case for the same head against the discharge intensity observed for $0.1 < H/P < 1$.

The present study explores the exponent b over head variation for increasing values of developed length ratio (*L/W*) against the fixed value of 1.5 used in literatures taking constant width of the channel for the case *H/P* greater than 0.1.

20.3 Methodology

20.3.1 Nonlinear Regression Analysis

Nonlinear regression is a curve-fitting tool intended to minimize the squared sum (SS) of the difference between observed data and the possible fit. However, it is an iterative or cyclical process where an initial guess of the parameters' values will have to be provided. In the successive iterations, parameter values are changed by a small amount and SS recalculated. This process is repeated until the changes in the parameter values result in minimizing the value of SS. Since in nonlinear regression, the second and higher derivatives are not zero, an iterative process is required to calculate the optimal parameter values. Nonlinear regression algorithm includes the Gauss–Newton, the Marquardt–Levenberg methods, gradient descent methods, etc. They all require an initial guess value of parameters and use them to estimate the parameters in an iterative process.

20.3.1.1 The Marquardt–Levenberg Algorithm

The Levenberg–Marquardt algorithm (LMA), also known as the damped least squares (DLS) method, is the most widely used method for solving nonlinear least square minimization problems. The LM algorithm interpolates between the Gauss–Newton algorithm (GNA) and the method of gradient descent. It is more robust than the GNA as it gives a solution even if the initial guess is very far from the final minimum. The LM algorithm provides a solution for nonlinear least squares minimization. This implies that the function has the following special form:

$$f(x) = \frac{1}{2}\sum_{j=1}^{j=m} {r_j}^2(x)$$

where $x = (x_1, x_2, \ldots x_n)$ is a vector, and each r_j is a function from $\Re$i to $\Re$. The r_j is referred to as a residual, and it is assumed that $m \geq n$. Now, f can be rewritten as $f(x) = \frac{1}{2}||r(x)||^2$. The derivatives can be written using the Jacobian matrix (J) of r with respect to x defined as $J(x) = \frac{\partial r_j}{\partial x_i}$ Where $1 \leq j \leq m$ and $1 \leq i \leq n$. Taking the sum of square deviation with respect to small increase δ and setting the residual to zero, we get:

$$(J^T J)\delta = J^T(y - f(\beta))$$

where f and y are vectors with ith component $f\,(x_{i,}\beta)$ and y_i.

For the general nonlinear case, we have:

$$\nabla f(x) = \sum_{j=1}^{m} r_j(x)\nabla r_j(x) = J(x)^T \mathrm{r}(\mathrm{x})$$

$$\nabla^2 f(x) = \sum_{j=1}^{m} r_j(x)\nabla^2 r_j(x) + J(x)^T J(x) \text{ and hence}$$

$$((J^T J) + \mu I)\delta = J^T(y - f(\beta))$$

where μ, is the dampening factor, I the identity matrix and δ is the small increment given to a parameter β.

The distinctive property of least squares problems is that from the given Jacobian matrix (J), the Hessian $\nabla^2 f(x)$ is easily found, and it is possible to approximate the r_j s through linear functions ($\nabla^2 r_j(x)$ are small) or the residuals ($r_{j(x)}$ themselves are small). The Hessian simply becomes$\nabla^2 f(x) = J(x)^T J(x)$. The non-negative dampening factor is updated after each iteration. A smaller value is used if the reduction in residual value is rapid, while if the iteration leads to an insufficient reduction in residual value, the dampening factor is increased.

20.3.1.2 ANOVA

ANOVA, also called analysis of variance, is a statistical tool to test differences between two or more means. ANOVA yields levels of variability within a regression model.

SST = SSM + SSE, where SS is a notation for the sum of squares and T, M, and E is a notation for total, model, and error, respectively.

R-squared or coefficient of determination is a statistical tool to measure data points' closeness to the fitted regression line. It is the percentage of variation in the response variable which can be explained through a linear model.

R^2 = (Explained variation)/(Total variation) and is between 0 and 100%:

- 0% value suggests that the model cannot explain the variability in the response data about its mean.
- 100% value suggests that the model can explain all the variability in the response data about its mean.
- Higher the R^2, the better the regression model fits your data.

20.4 Results and Discussion

A nonlinear regression analysis for *Q/LW*, where *Q* is the discharge, *L* the developed length, and *W* the width of the channel, was done from the experiments done by Kabiri-Samani and Javaheri (2012) to best fit the curve taking *Q/LW* as a function of a constant multiplied by head raised to an exponent variable *b*.

$$Q/LW = aH^b$$

The results of nonlinear regression analysis for various developed length are presented in Tables 20.1, 20.2, 20.3, 20.4, and 20.5. Table 20.6 summarizes the constant "*a*" and exponent "*b*" values for increasing *L/W* ratio for a fixed channel width of piano key weir (Fig. 20.2).

Table 20.1 **a** Parameter estimate for *L/W* = 2.5. **b** ANOVA table for *L/W* = 2.5

(a)

Parameter	Estimate	Std. error	95% confidence interval	
			Lower bound	Upper bound
a	0.678	0.035	0.594	0.761
b	1.193	0.024	1.135	1.251

(b)

Source	Sum of squares	df	Mean squares
Regression	466.881	2	233.440
Residual	0.156	7	0.022
Uncorrected total	467.037	9	
Corrected total	82.096	8	

Dependent variable: *Q/LW*[a]

$R^2 = 1 -$ (Residual sum of squares)/(corrected sum of squares) = 0.998

Table 20.2 **a** Parameter estimate for *L/W* = 4.0. **b** ANOVA table for *L/W* = 4.0

(a)

Parameter	Estimate	Std. error	95% confidence interval	
			Lower bound	Upper bound
a	0.512	0.025	0.453	0.570
b	1.186	0.022	1.134	1.238

(b)

Source	Sum of squares	df	Mean squares
Regression	299.359	2	149.679
Residual	0.082	7	0.012
Uncorrected total	299.440	9	
Corrected total	52.378	8	

Dependent variable: *Q/LW*[a]
$R^2 = 1 -$ (residual sum of squares)/(corrected sum of squares) = 0.998

Table 20.3 **a** Parameter estimate for *L/W* = 6.0. **b** ANOVA table for *L/W* = 6.0

(a)

Parameter	Estimate	Std. error	95% confidence interval	
			Lower bound	Upper bound
A	0.590	0.018	0.545	0.635
B	1.015	0.015	0.978	1.052

(b)

Source	Sum of squares	df	Mean squares
Regression	148.421	2	74.210
Residual	0.015	6	0.002
Uncorrected total	148.435	8	
Corrected total	14.909	7	

Dependent variable: *Q/LW*[a]
$R^2 = 1 -$ (residual sum of squares)/(corrected sum of squares) = 0.999

20.5 Conclusion

The data is taken from literature, and as such, the same limitations apply here also. The results as such are subject to the present tests limitations $h > 30$ mm, $0.1 \leq H/P \leq 0.6$, $2.5 \leq L/W \leq 7$, $1 \leq B/P \leq 2.5$, $0.33 \leq W_i/W_o \leq 1.22$, $0 \leq B_i/B \leq 0.26$, $0 \leq B_o/B \leq 0.26$, and $H_d/H \leq 0.6$

The nonlinear regression analysis shows that as we bring more and more fold in the weir geometry to increase the crest length for more significant discharge release through PKW in a constant width of channel width, the exponent (b) of head (h) decreases from the analytical value of 1.5 to almost 0.9.

Table 20.4 **a** Parameter estimate for *L/W* = 7.0. **b** ANOVA table for *L/W* = 7.0

(a)

Parameter	Estimate	Std. error	95% confidence interval	
			Lower bound	Upper bound
a	0.584	0.021	0.535	0.633
b	0.975	0.019	0.932	1.018

(b)

Source	Sum of squares	df	Mean squares
Regression	109.120	2	54.560
Residual	0.030	8	0.004
Uncorrected total	109.150	10	
Corrected total	17.117	9	

Dependent variable: *Q/LW*[a]
$R^2 = 1 -$ (Residual sum of squares)/(corrected sum of squares) = 0.998

Table 20.5 **a** Parameter estimate for *L/W* = 8.5. **b** ANOVA table for *L/W* = 8.5

(a)

Parameter	Estimate	Std. error	95% confidence interval	
			Lower bound	Upper bound
a	0.616	0.026	0.556	0.676
B	0.879	0.022	0.830	0.929

(b)

Source	Sum of squares	df	Mean squares
Regression	90.386	2	45.193
Residual	0.036	8	0.004
Uncorrected total	90.422	10	
Corrected total	11.644	9	

Dependent variable: *Q/LW*[a]
$R^2 = 1 -$ (residual sum of squares)/(corrected sum of squares) = 0.997

For $H/P < 0.1$, we have observed that exponent b drops to 0.75 though the discharge intensity that can be passed through PKW increases for the same head as opposed to $0.1 < H/P < 0.6$. This may be likely due to the effect of surface tension force which was excluded from analysis by Kabiri-Samani and Javaheri (2012) by taking $H >$ 30 mm.

Since in confined channel width, as we increase the developed length ratio, the width of inlet and width of outlet will automatically decrease. To accommodate this, we have to increase the overhangs, which are made structurally stable by increasing the weir height. The entire analysis was carried out about a single parameter, i.e., the developed length ratio (*L/W*), since all the geometrical parameters are interrelated.

Table 20.6 Values of "a" and "b" for increasing L/W ratio

S. no.	L/W	a	B	R^2 value
1	2.5	0.678	1.193	0.998
2	4.0	0.512	1.186	0.998
3	6.0	0.590	1.015	0.999
4	7.0	0.584	0.975	0.998
5	8.5	0.654	0.879	0.997

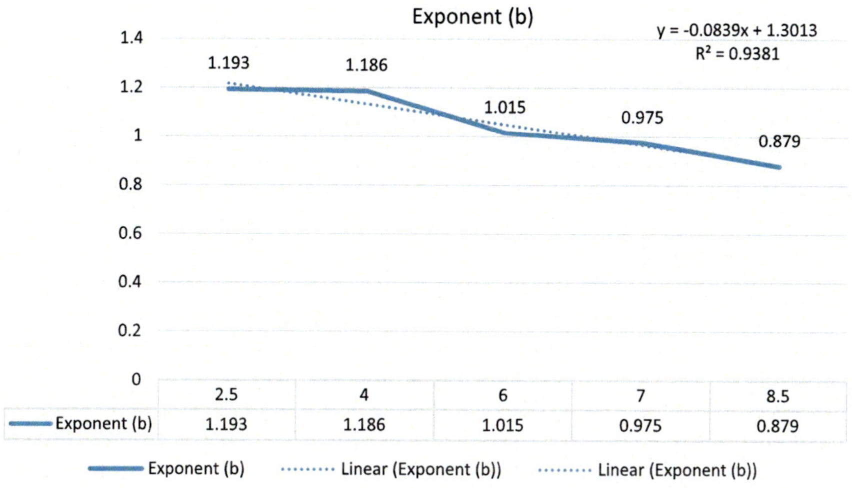

Fig. 20.2 Change in exponent "*b*" of head over increasing *L/W* value for flow over PKW in a channel of constant width

References

Anderson RM (2011) Piano key weir head discharge relationships, Utah State University

Anderson R, Tullis B (2011). Influence of piano key weir geometry on discharge. In: Proceedings of international conference labyrinth and piano key weirs Liège B

Anderson R, Tullis B (2012) Piano key weir hydraulics and labyrinth weir comparison. J Irrig Drain Eng 139(3):246–253

Kabiri-Samani A, Javaheri A (2012) Discharge coefficients for free and submerged flow over piano key weirs. J Hydraul Res 50(1):114–120

Laugier F, Blancher B (2011) Influence of structural thickness of sidewalls on PKW spillway discharge capacity

Laugier F, Vermeulen J, Pralong J (2012) Achievement of new innovative labyrinth piano key weir spillways (PKW). In: Proceedings of piano key weir for in-stream storage and dam safety (pKwIsD-2012), New Delhi, pp 25–42

Leite Ribeiro M, Bieri M, Boillat J-L, Schleiss A, Delorme F, Laugier F (2009) Hydraulic capacity improvement of existing spillways-design of a piano key weirs. In: Proceedings (on CD) of the 23rd congress of the international commission on large dams CIGB-ICOLD

Leite Ribeiro M, Bieri M, Boillat J-L, Schleiss A, Singhal G, Sharma N (2011) Discharge capacity of piano key weirs. J Hydraul Eng 138(2):199–203

Lempérière F, Ouamane A (2003) The Piano Keys weir: a new cost-effective solution for spillways. Int J Hydropower Dams 10(5):144–149

Lempérière F, Vigny J (2011) General comments on labyrinth and piano keys weirs—the future. In: Labyrinth and piano key weirs–PKW 2011, pp 289–294

Mehboudi A, Attari J, Hosseini S (2016) Experimental study of discharge coefficient for trapezoidal piano key weirs. Flow Meas Instrum 50:65–72

Pralong J, Vermeulen J, Blancher B, Laugier F, Erpicum S, Machiels O, Pirotton M, Boillat J-L, Leite Ribeiro M, Schleiss A (2011) A naming convention for the piano key weirs geometrical parameters. In: Labyrinth and piano key weirs, pp 271–278

Sharma N, Tiwari H (2013) Experimental study on vertical velocity and submergence depth near piano key weir. In: Labyrinth and piano key weirs II-PKW, pp 93–100

Tiwari H, Sharma N (2015) Developments to improve hydraulic competence of spillways. Aquatic Proc 4:841–846

Tiwari H, Sharma N (2017) Empirical and Mathematical modeling of head and discharge over piano key weir. In: Development of water resources in India. Springer, pp 341–354

Chapter 21
Grey Water Characterization and Its Management

Sarosh Alam Ghausi and Mohd Muzzammil

Abstract Water, the driving force of nature, is the life sustaining resource present on the earth. Depleting groundwater tables and wells running dry has forced the researchers and policymakers to come up with an efficient solution to optimize water use and save water. With rainfall behaviour being uncertain and diminishing surface water sources, there is a critical need to look out for alternate water source. In this scenario, grey water use provides an efficient option. Grey water is the waste water generated from showers, laundry washing, washbasins and sinks. It does not come in contact with organic impurities which result in lower BOD as compared to black water (water generated from toilet flushing). The basic idea is to separate grey water from domestic sewage. Depending on the type of grey water and its level of treatment, it can be reused for various purposes like irrigation, flushing, floor washing, watering gardens, automobile washing, etc. Studies have suggested that recycling grey water can save up to 70% of fresh water consumption. To control the growing water imbalance, grey water use has huge potential to be used as sustainable alternate resource. Present study characterizes different types of grey water in terms of water quality parameters and suggests a low-cost treatment system for its further use.

Keywords Grey water · Characterization · Low-cost treatment system · Water quality parameters

S. A. Ghausi (✉)
Civil Engineering Department, IIT BOMBAY, Mumbai 400076, India

M. Muzzammil
Civil Engineering Department, Z.H. College of Engineering & Tech, AMU, Aligarh, U.P. 202002, India

R. Jha et al. (eds.), *Water Resources Management and Reservoir Operation*, Water Science and Technology Library 107,
https://doi.org/10.1007/978-3-030-79400-2_21

21.1 Introduction

As per United Nations department of economic and social affairs (UNDESA) report, almost one-fifth of world that is around 1.2 billion people live in areas of physical water scarcity with another 1.6 billion people approaching this situation (FAO 2007). It is projected that by 2025 about 3.5 billion people will live in water stressed countries (UNESCO 2012). More than one in every six people in the world do not have access to potable water. Around 700 million people in 43 countries suffer from water scarcity (Global Water institute 2013). A third of the world's biggest groundwater systems are already in distress (Richey et al. 2016). There are several other studies depicting current water crisis and thus it is evidently clear to look out for new advances and technology for water conservation and its management. There is an urgent need to critically look into the alternate sources of water. Talking about alternate sources, rainwater is the first option coming into the mind but the major problem with rain is that it is uncertain and non-homogeneous and could not fulfil the constant demand of water. Expanding industrialization has highly reduced the surface water source. Problem with desalination is the mineral decomposition of potable water, thus in this scenario grey water use is a highly efficient option to be used as an alternate resource. While rainwater harvesting and Integrated Water Resource Management (IWRM) are water conservation measures, grey water use is an important undeveloped technology.

Grey water is the waste water generated from showers, laundry washing, wash-basins and sinks. Depending on the type of grey water and its level of treatment, it can be reused for various purposes like irrigation, flushing, floor washing, watering gardens, automobile washing, etc. Studies have suggested that recycling grey water can save up to 70% of fresh water consumption (Alam et al. 2012; Ghausi et al. 2017). Since it does not come in contact with human waste, its water quality parameters like BOD are much lesser than usual sewage and it require much lower degree of treatment. Thus, grey water use is important because it restricts water demand and reduces stress on treatment system. Table 21.1 clearly depicts the difference between water quality parameters in grey and sewage water (Rana et al. 2014). It can be observed that biochemical oxygen demand (BOD) and chemical oxygen demand (COD) are much lower for grey water as compared to sewage water.

Table 21.1 Comparison of water quality parameters of grey and sewage water

Parameter	Unit	Grey water	Sewage water
PH	–	6.4–8.1	5.5–8.5
Suspended solids	Mg/l	40–340	100–350
BOD_5	Mg/l	40–50	100–300
COD	Mg/l	100–300	175–600
Nitrogen (total)	Mg/l	2–23	30–60
Phosphorus	Mg/l	0.1–0.8	7–20
Turbidity	NTU	15–270	–

Grey water can be used for various purposes that include irrigating plants, watering fields, toilet flushing, floor washing, car washing, recharging of dry ponds, etc. Separate grey water lines from sewer lines in domestic households could be a very good concept in designing of new cities. Grey water use is a huge success in USA with around 20 states using it. However, there are conflicts to permit its use for irrigation.

21.2 Characterization of Grey Water

The sources of grey water generated from domestic households mainly include water from laundry, washbasin and bathing. The water generated from kitchen is considered as dark grey water due to comparatively high content of organic impurities and hence not taken up in grey water characterization. Characterization is done on basis of various water quality parameters that are Conductivity, alkalinity, total hardness, BOD_5, COD, chloride, turbidity and PH.

Water is characterized as

1. Laundry grey water
2. Bathing grey water
3. Washbasin grey water

Sample collection for laundry grey water was done by collecting the grey water from drain of a washing machine. For washbasin, grey water sample was collected by storing the water from washbasin pipe into a large container. Soap used was an antiseptic soap, whereas for bathing, grey water was collected by storing the water before it reaches the drain, and the soap used was a commercial bathing soap. Different water quality parameters of these samples are shown in Table 21.2.

From the following data broadly it can be inferred that grey water generated from laundry is most polluted while that in washbasin is least polluted. It was observed that laundry water was more turbid and dark grey in colour with moderate smell whereas washbasin and bathing sample were greyish white and have slight smell. Grey water in washbasin sample seems to become clearer with time due to sedimentation whereas for laundry sample it requires skimming of foam and lather from the top surface. PH of all the three samples was found to be almost with in neutral range; however, laundry sample seems to have slightly alkaline nature as it also shows the highest alkalinity value. BOD_5 was maximum for laundry sample and minimum for washbasin sample and so was the case of COD. The COD for laundry sample was much larger than for washbasin and bathing sample; however, it should be noted that COD value is directly related to dilution of sample. Concentrated grey water sample may lead to much higher value of COD (>1000).

Following graph depicts the comparison of different parameters of grey water samples.

Table 21.2 Water quality parameters of grey water

	Laundry	Washbasin	Bathing
Alklanity (mg/l)	668	384	410
Hardness (mg/l)	270	332	254
Chloride (mg/l)	236	92.4	60
BOD_5 (mg/l)	49.12	30	35
COD (mg/l)	960	360	380
Conductivity (μS/cm)	1400	1127	1275
Turbidity (NTU)	165	93	108
PH	8.5	7.41	7.76
TDS (mg/l)	1475	1073	1167
Colour	Dark grey	Greyish white	Greyish white
Odour	Moderate	Slight	Slight

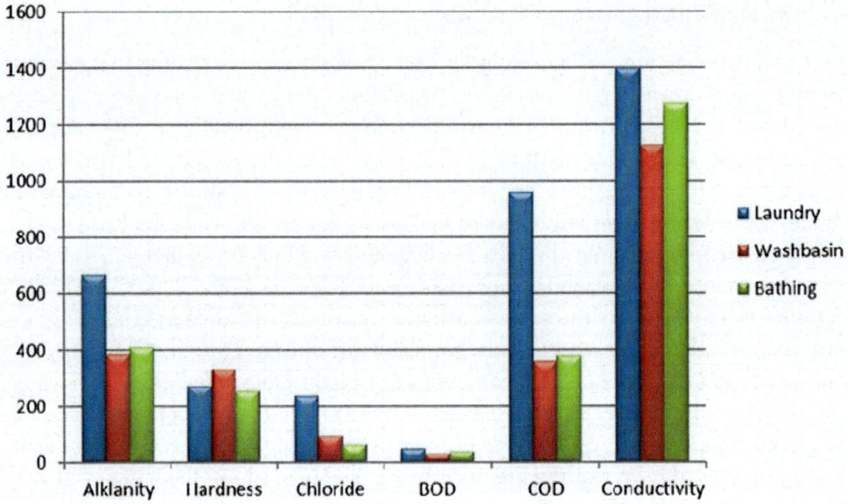

21.3 Treatment of Grey Water

The main objective in treatment of grey water is the removal of soapy colloidal impurities which are responsible for the turbidity in it. Sand bed filters are often used to remove the particles for grey water. Study proposes a model of low-cost treatment tank for treating grey water.

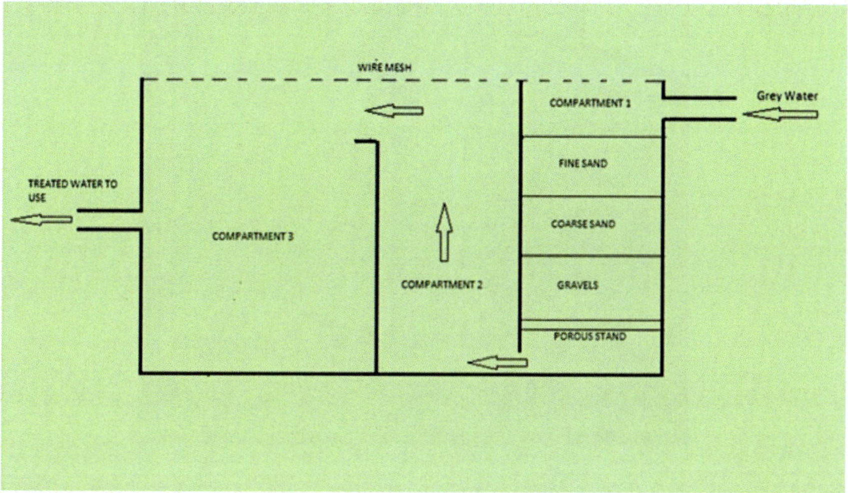

The treatment tank is subdivided into three compartments. Compartment 1 is provided with layers of fine sand, coarse sand and gravels (10 mm) resting on a porous stand. All these materials are cheaply and readily available. Grey water is allowed to pass through this compartment and then it enters compartment 2 where it rises against the gravity. Particles if any will settle down in this compartment. In this compartment, some disinfectant may also be added if required or bleaching can be done to prevent odour problems. Some flocculating agent may also be added if colloidal impurities still persist. From compartment 2, water is allowed to fall in compartment 3 increasing the turbulence and hence more air–water interfaces which will subsequently result in increased aeration.

Experiments were performed to check the efficiency of this tank, and similar conditions were simulated and tested for compartment 1.

Grey water samples of laundry, washbasin and bathing were passed through this filter and there water quality was checked before and after passing.

21.3.1 Laundry Sample

Comparison between water quality parameters of direct and filtered sample is made clear in Table 21.3.

The desirable and permissible limits for drinking water standards are as per IS 10500: 1991. For drinking water standards, there is no standard value for BOD and COD mentioned but BOD <30 mg/l for inland surface water and <100 mg/l for irrigation water and COD <250 mg/l. It can be inferred from the table and the graph that after filtration the water quality of grey water is highly improved. BOD value of filtered sample is less than 10. There is huge decrease in total dissolved solids which came into the permissible range of less than 500. Also, the turbidity is highly reduced and the water becomes clear. All the water quality parameters are with in permissible limits (Kumar and Avinash 2012).

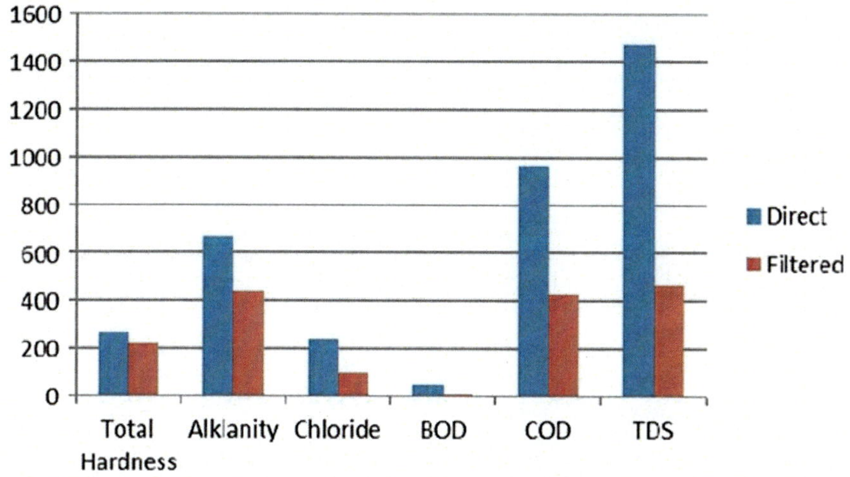

Underlying figure shows the direct laundry grey water sample (left) and filtered sample (right).

Table 21.3 Water quality of laundry grey water and filtered water

	Direct sample	Filtered sample	desirable limits	Permissible limits
Total hardness (mg/l)	270	220	300	600
Alkalinity (mg/l)	668	440	200	600
Chloride (mg/l)	236	98	250	1000
BOD_5 (mg/l)	49.12	9.7	30	100
COD (mg/l)	960	430	–	–
TDS	1475	470	500	2000
Turbidity (NTU)	165	7	10	10
PH	8.5	7.66	6.5–8.5	6.5–8.5
Colour	Dark grey	Colourless		
Odour	Moderate	No		

Table 21.4 Water quality of bathing grey water and filtered water

	Direct sample	Filtered sample	Desirable limits	Permissible limits
Total hardness	254	178	300	600
Alkalinity	410	326	200	600
Chloride	60	50	250	1000
BOD	35	8.9	30	100
COD	380	244	–	–
TDS	1167	443	500	2000
Turbidity	108	5	10	10
PH	7.76	7.48	6.5–8.5	6.5–8.5
Colour	Greyish white	Colourless		
Odour	Slight	No		

21.3.2 *Bathing Sample*

Comparison between water quality parameters of direct and filtered sample is made clear in Table 21.4.

From Table 21.4 also, similar behaviour can be inferred as in Table 21.3. There is a considerable decrease in BOD_5, COD, turbidity and TDS of the sample.

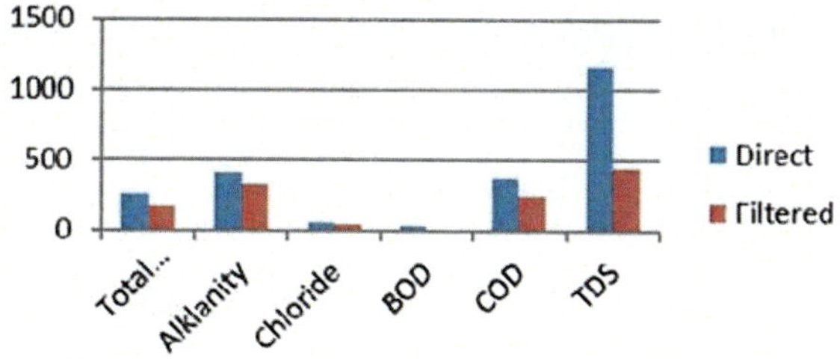

Following graph shows how the water quality of filtered sample is enhanced. Underlying figure shows direct bathing grey water sample (left) and filtered sample (right) (Table 21.5).

Table 21.5 Water quality of direct and filtered washbasin sample

	Direct sample	Filtered sample	Desirable limits	Permissible limits
Total hardness (mg/l)	332	268	300	600
Alkalinity (mg/l)	384	310	200	600
Chloride (mg/l)	92.4	72.8	250	1000
BOD_5 (mg/l)	30	6.55	30	100
COD (mg/l)	360	220	–	–
TDS (mg/l)	1073	411	500	2000
Turbidity (NTU)	93	3	10	10
PH	7.41	7.38	6.5–8.5	6.5–8.5
Colour	Greyish white	Colourless		
Odour	Slight	No		

21.3.3 Washbasin Sample

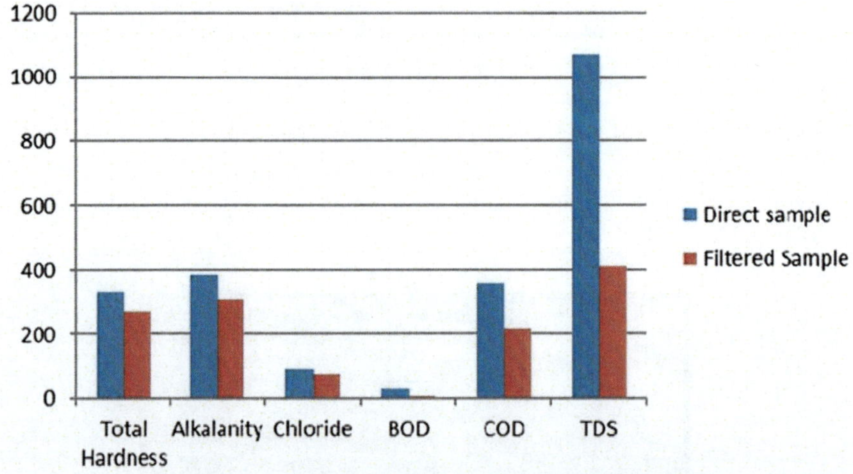

Washbasin sample is the purest form of grey water and on further treatment it can be observed that BOD_5 has decreased to mere 6 mg/l whereas turbidity has fallen to 3 NTU making water almost perfectly clear. Underlying figure shows the washbasin grey water sample (left) and filtered sample (right).

21.4 Conclusion

Present study focuses on the use of grey water as an alternate resource to solve the increasing problem of water scarcity. It characterizes grey water on the basis of source of generation from domestic households as laundry grey water, washbasin grey water and bathing grey water. On basis of water quality experiments done, it was concluded that washbasin grey water is the purest form of grey water. A model of low-cost treatment tank was also proposed. Experiment results showed that there is vast improvement in water quality of grey water by passing it through the proposed filter. With growing imbalance between water demand and water, grey water has a huge potential to serve as an alternate water resource and it will not be wrong to say that it is high time and grey water use is the need of the hour. It is only the matter of time of how early we realize its importance.

References

Alam J, Muzzamil M (2012) Grey water use: a need of hour. India water week, Ministry of water Resources, New Delhi

General standards for discharge of environmental pollutants (Schedule VI) (1986) Environment protection rules

Ghausi SA, Muzzammil M, Alam J (2017) Grey water use as an alternate resource. In: Proceedings of 37th IAHR world congress held at Kualalumpur, Malaysia w.e.f 13th to 18th Aug, pp 4722–4729

Indian standard for drinking water—Specifications IS 10500 (1991)

Kumar M, Puri A (2012) A review of permissible limits of drinking water. Indian J Occupational Environ Med

Rana K, Shah M, Upadhyay A (2014) Integrated approach towards grey water management. Int J Eng Sci Res Technol 3(1):239–242

Richey AS, Thomas BF (2016) Quantifying renewable ground water stress with GRACE

Sarkar P, Sharma B, Malik U (2014) Energy generation from grey water in high rised buildings: the case of India. Renew Energy 69:284–289

Chapter 22
Intelligent Operation of Hirakud Reservoir Using Metaheuristic Techniques (PSO and TLBO)

Pooja Patnaik and Prakash Ch. Swain

Abstract In water resources engineering, the real-life problems are mostly involved with the nonlinear formulations. It is becoming a difficult task for large-scale nonlinear optimization problems to obtain the optimal solutions. Since various conflicting demands such as irrigation, power production, industrial water supply, municipal water supply, etc., should be satisfied with water available in reservoir, the optimal operating policy for multipurpose reservoir is a necessity. In this study, the metaheuristic techniques like particle swarm optimization (PSO) and teaching–learning-based optimization (TLBO) approaches are developed to overcome the limitations of conventional techniques. The main objective of this paper is to develop a policy for optimizing the total release of water for irrigation, power generation and industries during non-monsoon period with the case study of the multipurpose Hirakud reservoir of Odisha state in India. The TLBO and PSO models are implemented in MATLAB, and the developed programs are executed on a 4 GB RAM, 64-bit operating system, Quad core processor, and it is observed that the computational time required for iteration of TLBO and PSO is 10 h 33 min and 11 h 37 min, respectively, whereas in PARAMsavak supercomputer of 64 GB RAM, Ubuntu-operating system, 28 logical core processor the computational time required for 300 iterations of TLBO and PSO are 1 h 22 min and 1 h 54 min, respectively. Therefore, supercomputer is used for running the model as it saves computational time. On comparing the policy developed by PSO and TLBO methods for reservoir operation problems, it is observed that the release of water for irrigation, power generation, and industrial purpose are more by using TLBO than that of the release by other techniques, keeping strict surveillance on arriving at the dead storage level of the reservoir at the end of the non-monsoon period.

Keywords Reservoir operation · Particle swarm optimization · Teaching–learning-based optimization · Reservoir management · Hirakud dam

P. Patnaik (✉) · P. Ch. Swain
Civil Engineering Department, VSSUT, Burla 768018, India

R. Jha et al. (eds.), *Water Resources Management and Reservoir Operation*,
Water Science and Technology Library 107,
https://doi.org/10.1007/978-3-030-79400-2_22

22.1 Introduction

In India, there are several single and multipurpose reservoir projects. The reservoir water is often used to meet the requirements for various purposes such as municipal and industrial water supply, water supply for irrigation, hydropower generation, etc. Nowadays, it is a complex task for the optimal planning and appropriate management of multipurpose reservoir. For optimizing the reservoir operation problems, formerly many strategies were developed. Based on the idea of characteristics of reservoir, statistics, constraints and objective characteristic, the selection of the optimization techniques are done. Because of each problem contains its precise physical and operating characteristics, there is no standard algorithm available which can be relevant to all problems since four decades reservoir operation is one of the important research area for water resources field. Many mathematical models have been purposed during this period. Some of the research works are described below. Eberhart and Kennedy (1995) proposed PSO for continuous nonlinear functions. They discussed two paradigms that put in force the concept, one was globally orientated (gbest), and other was locally oriented (lbest), accompanied by way of effects received from applications and tests upon which the paradigms have been shown to perform correctly. Reddy and Kumar (2009) used elitist-mutated multi-objective particle swarm optimization (EM-MOPSO) method for integrated water resources management. Their objective was to minimize flood risk, maximize hydropower production, and minimize irrigation deficits in a year, subject to various physical and technical constraints. Swain and Nanduri (2009) used neuro-fuzzy inference system for finding the reservoir management policy of a multipurpose reservoir system, i.e., Hirakud reservoir. This model was developed for forecasting inflows into the reservoir and operation of the reservoir during the monsoon period. Recently, a new metaheuristic technique that is teaching–learning-based optimization (TLBO) has been recently developed and is becoming more popular for solving the optimization problems (Rao and Teja 2015) as it uses only common controlling parameters such as number of generations and population size and no specified control parameters. It causes difficulties and complexity when dealing with the reservoir operation. So a proper and effective operation policy should be done which can deal with all associated problem and produce a proper standard policy for the reservoir.

22.2 Study Area and Data Details

The Hirakud reservoir is situated in Orissa state, India. The dam is built across river Mahanadi at about15 km upstream of Sambalpur town in the state of Orissa and dam is in operation since 1957. The geographical location of Hirakud dam in global grid is at latitude 21°32″ N and longitude 83°52″ E. The location map of Hirakud reservoir is shown in Fig. 22.1. Hirakud reservoir is one of the biggest artificial lake in Asia with reservoir spread of 743 km^2 at full reservoir level. The water surface area at

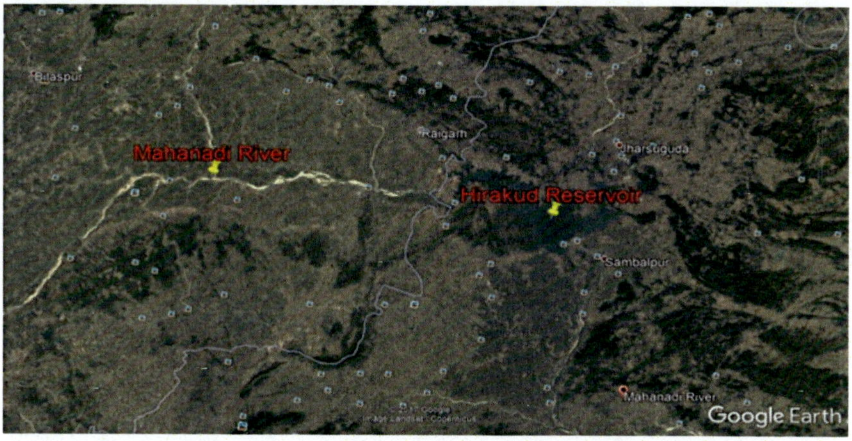

Fig. 22.1 Satellite image of Hirakud reservoir. (*Source* Google Earth)

dead storage level is around 274 km^2. The reservoir has gross, live, and dead storage capacities of 7189, 5375, and 1814 million cubic meter (Mm^3), respectively. The reservoir elevations at gross storage and dead storage are 192.024 m and 179.830 m, respectively.

The Hirakud reservoir system serves multiple purposes such as flood control, irrigation, and hydropower. In monsoon season, the first priority of the reservoir is to minimize the adverse effects of floods. From the analysis of past historical data and practical experience, the Hirakud dam authority follows some guidelines to maintain the reservoir levels at different time periods to protect the downstream region from flooding. Thus, when formulating mathematical model for reservoir operation, these flood control restrictions need to be satisfied at various time periods. Then, the project provides water for irrigation in the districts of Sambalpur, Bargarh, Bolangir, and Subarnapur for an area of 155,635 ha in Kharif season (June–October) and 108,385 ha in Rabi season (November–April). The water released through the power house irrigates the Mahanadi Delta area of 436,000 ha. The project provides hydropower generation through two power houses, namely Burla and Chiplima. Data of 27 years are taken on ten daily basis from 1990 to 2016. Flow data are collected from odisha water planning organization, Secha Sadan, Bhubaneswar; office of Chief Engineer-cum-Basin Manager, Upper Mahanadi, Burla; office of the executive engineer, Hirakud dam circle, Burla; master control room, Burla; central water commission, Burla and Bhubaneswar; Indian Meteorological Department, Hirakud and Bhubaneswar; Survey of India, Bhubaneswar. Data are processed before feeding to the models.

22.3 Methodology

Optimization techniques are used in every sector such as water demand management, construction program management, industrial management, power system control, etc. Main objective of using optimization is determining variables and determining the optimum result. Word 'optimization' refers to perfection. It defines as choosing best from set available. It is a process of making perfection and more effective results. It deals with the best solution in a simple way. With progression of time, new methods are introduced in optimization from classical methods to evolutionary algorithm. Particle swarm optimization and teaching–learning-based optimization are used. They are broadly discussed below.

22.3.1 Particle Swarm Optimization Algorithm

The practical swarm optimization (PSO) is one of the metaheuristic methods that are widely used in water resource management. It was initially proposed by Eberhart and Kennedy (1995). It has been inspired by the social behavior of animals like bird flocking, fish schooling, and insect swarming. The PSO used the idea that social sharing of information among members may have some evolutionary advantage. PSO has been applied to many real-world problems. A standard PSO algorithm was initialized with a population (swarm) of random potential solutions (particles). Each particle iteratively moves across the search space and was attracted to the position of the best fitness historically achieved by the particle itself (local best) and by the best among the neighbors of the particle (global best). In PSO, instead of using more traditional genetic operators, each particle adjusts its flying according to its own flying experience and its companions' flying experience.

Assume the place of the goal is in the M dimension. Individual particle is denoted by, i.e., $X_j = \left(x_{j1}, x_{j2}, \ldots x_{jM}\right)^T$. The velocity of each particle is represented by $V_j = \left(V_{j1}, V_{j2}, \ldots V_{jM}\right)^T$. The position of each particle with personal best position is represented as $PB_{id}^n = (pb_{i1}, pb_{i2}, \ldots pb_{iD})^T$. The global best value of individuals is denoted as $PB_{gd}^n = \left(pb_{g1}, pb_{g2}, \ldots pb_{gD}\right)^T$.

The following equations give the updating rules for velocity and position of the particle:

$$V_{id}^{n+1} = \chi\left[\omega V_{id}^n + c_1 r_1^n \frac{\left(PB_{id}^n - X\right)}{\Delta t} + c_2 r_2^n \frac{\left(PB_{gd}^n - X\right)}{\Delta t}\right] \tag{22.1}$$

$$X_{id}^{n+1} = X_{id}^n + (\Delta t) V_{id}^{n+1} \tag{22.2}$$

where χ is the constriction coefficient, x is the inertial weight, c_1 and c_2 are the cognitive and social parameters of PSO, $d = 1, 2, \ldots, D$, the index for decision variables, i = 1,2,..., NS, the index for swarm population, NS is the size of the swarm, g is the index of best particle among all particles in the population, n is the generation number, Δt is time step (considered as unity) and r_1 and r_2 are uniformly generated random numbers between 0 and 1. By the effect of these two equations, individual particle changes its velocity and position with respect to its previous best, the best position of its neighbor. For getting the faster convergence, proper tuning c_1 and c_2 is needed. The inertia weight is used to control the effect of previous velocity in new one. The larger value of inertia weight refers to the global exploration, whereas the smaller value of inertia weight refers to the local exploration. So choosing the value of the inertia weight is very important that will help to reduce the number of iterations.

22.3.2 *Teaching–Learning-Based Optimization*

The TLBO algorithm is a teaching–learning process inspired algorithm proposed by Rao et al. (2011; 2012a, b) and Rao and Savsani (2012) based on the effect of influence of a teacher on the output of learners in a class. The algorithm describes two basic modes of the learning: (i) through teacher (known as teacher phase) and (ii) through interaction with the other learners (known as learner phase). In this optimization algorithm, a group of learners is considered as population, and different subjects offered to the learners are considered as different design variables of the optimization problem and a learner's result is analogous to the 'fitness' value of the optimization problem. The best solution in the entire population is considered as the teacher. The design variables are actually the parameters involved in the objective function of the given optimization problem, and the best solution is the best value of the objective function. The working of TLBO is divided into two parts, 'Teacher phase' and 'Learner phase.' Working of both the phases is explained below.

22.3.2.1 Teacher Phase

It is the first part of the algorithm where learners learn through the teacher. During this phase, a teacher tries to increase the mean result of the class in the subject taught by him or her depending on his or her capability. At any iteration i, assume that there are 'm' number of subjects (i.e., design variables), 'n' number of learners (i.e., population size, $k = 1, 2, \ldots n$) and $M_{j,i}$ be the mean result of the learners in a particular subject 'j' ($j = 1, 2, \ldots m$) The best overall result $X_{\text{total-kbest},i}$ considering all the subjects together obtained in the entire population of learners can be considered as the result of best learner kbest. However, as the teacher is usually considered as a highly learned person who trains learners so that they can have better results, the best learner identified is considered by the algorithm as the teacher. The difference between the existing mean result of each subject and the corresponding result of the teacher for each subject is given by,

$$\text{Difference}_{\text{Mean}_{j,k,i}} = r_i\left(X_{j,\text{kbest},i} - T_F M_{j,i}\right) \tag{22.3}$$

where $X_{j,\text{kbest},i}$ is the result of the best learner in subject j; T_F is the teaching factor which decides the value of mean to be changed, and r_i is the random number in range [0,1]. Value of T_F can be either 1 or 2. The value of T_F is decided randomly with equal probability as,

$$T_F = \text{round}[1 + \text{rand}(0, 1)\{2 - 1\}] \tag{22.4}$$

T_F is not a parameter of the TLBO algorithm. The value of T_F is not given as an input to the algorithm, and its value is randomly decided by the algorithm using Eq. (22.4). After conducting a number of experiments on many benchmark functions, it is concluded that the algorithm performs better if the value of T_F is between 1 and 2. However, the algorithm is found to perform much better if the value of TF is either 1 or 2, and hence to simplify the algorithm, the teaching factor is suggested to take either 1 or 2 depending on the rounding up criteria given by Eq. (22.4). Based on the Difference_Mean$_{j,k,i}$, the existing solution is updated in the teacher phase according to the following expression.

$$X'_{j,k,i} = X_{j,k,i} + \text{Difference\_Mean}_{j,k,i} \tag{22.5}$$

where $X'_{j,k,i}$ is the updated value of $X_{j,k,i}$. $X'_{j,k,i}$ is accepted if it gives better function value. All the accepted function values at the end of the teacher phase are maintained, and these values become the input to the learner phase. The learner phase depends upon the teacher phase.

22.3.2.2 Learner Phase

It is the second part of the algorithm where learners increase their knowledge by interacting among themselves. A learner interacts randomly with other learners for enhancing his or her knowledge. A learner learns new things if the other learner has more knowledge than him or her. Considering a population size of '*n*,' the learning phenomenon of this phase is explained below.

Randomly select two learners P and Q such that $X'_{\text{total}-P,i} \neq X'_{\text{total}-Q,i}$ (where, $X'_{\text{total}-P,i}$ and $X'_{\text{total}-Q,i}$ are the updated function values of $X'_{\text{total}-P,i}$ and $X'_{\text{total}-Q,i}$ of P and Q, respectively, at the end of teacher phase)

$$X''_{j,P,i} = X'_{j,P,i} + r_i\left(X'_{j,P,i} - X'_{j,Q,i}\right), \quad \text{if} \quad X'_{\text{total}-P,i} < X'_{\text{total}-Q,i} \tag{22.6}$$

$$X''_{j,P,i} = X'_{j,P,i} + r_i\left(X'_{j,Q,i} - X'_{j,P,i}\right), \quad \text{if} \quad X'_{\text{total}-Q,i} < X'_{\text{total}-P,i} \tag{22.7}$$

$X''_{j,P,i}$ is accepted if it gives a better function value.

Equations (22.6) and (22.7) are for minimization problems. In the case of maximization problems, Eqs. (22.8) and (22.9) are used.

$$X''_{j,P,i} = X'_{j,P,i} + r_i\left(X'_{j,P,i} - X'_{j,Q,i}\right), \quad \text{if} \quad X'_{\text{total}-Q,i} < X'_{\text{total}-P,i} \tag{22.8}$$

$$X''_{j,P,i} = X'_{j,P,i} + r_i\left(X'_{j,Q,i} - X'_{j,P,i}\right), \quad \text{if} \quad X'_{\text{total}-P,i} < X'_{\text{total}-Q,i} \tag{22.9}$$

Teaching–learning-based optimization (TLBO) is a population-based algorithm which simulates the teaching–learning process of the class room. This algorithm requires only the common control parameters such as the population size and the number of generations and does not require any algorithm-specific control parameters.

22.4 Reservoir Operation Model

Main aim of any water resources project is to get minimum irrigation deficit with maximum hydropower production. These two objectives are conflicting to each other. For minimizing irrigation deficit, more release of water is required, and for hydropower generation, more water should be stored for maintaining high water level for production of more water levels. But nowadays, industrialization is one of the most important factor to improve the economic growth of a country. So the supply of water to the industries cannot be over ruled. Hence, the demand and release of water for industrial purpose are another parameter for the specific multipurpose reservoir operating model.

- **For irrigation release operation**

The square deficit should be minimized

$$\text{Minimize } Z = \sum_{t-1}^{T} (\text{IRD} - \text{IR})^2 \tag{22.10}$$

where Z is squared deviation of irrigation demand and releases; IRD is irrigation demands in period $t(10^6)$ m^3; IR is water release in period $t(10^6)$ m^3 at $t = 1,2,3,\ldots$.

- **For hydropower production operation**

  Maximizing the production of energy

$$\text{Maximize } E = \sum_{t-1}^{T} p(P)_t(H)_t \tag{22.11}$$

where E is Energy produced (KW); p is power production coefficient; P_t is release made to river bed turbine in period (10^6) m^3; H_t is net heads available (m).

- **For industrial release operation**

The square deficit should be minimized

$$\text{Minimize } F = \sum_{t-1}^{T} (\text{IND} - \text{INR})^2 \tag{22.12}$$

where F is squared deviation of irrigation demand and releases; IND is industrial demands in period $t(10^6)$ m^3; INR is water release in period $t(10^6)$ m^3 at $t = 1,2,3,\ldots$.

- **For combining irrigation release and hydropower generation operation**

According to the basis of priority of the objective of the multipurpose reservoir, the final fitness function is

$$\text{Minimize } Z = cf_1 \sum_{t-1}^{T} \left(\frac{\text{IRD} - \text{IR}}{\text{IRD}} \right)^2 + cf_2 \sum_{t-1}^{T} \frac{E_{\max} - P(P)_t (H)_t}{E_{\max}}$$
$$+ cf_3 \sum_{t-1}^{T} (\text{IND} - \text{INR})^2 \tag{22.13}$$

where cf_1, cf_2, and cf_3 is constants are used according to the priority; IRD is irrigation demands in period $t(10^6)$ m^3; IR is water release in period $t(10^6)$ m^3; $E_{\max}$ is maximum energy produced (KW); p is power production coefficient; P_t is release made to river bed turbine in period (10^6) m^3; H_t is net heads available (m); IND is industrial demands in period $t(10^6)$ m^3; INR is water release in period $t(10^6)$ m^3.

22.4.1 Constraints for Irrigation Release, Hydropower Generation and Industrial Release

(a) Continuity constraints of the reservoir

$$S(t+1) = \text{St}(t) + \text{In}(t) - P(t) - \text{IRR}(t) - \text{Evp}(t) - \text{Ovf}(t) - \text{IND}(t) \tag{22.14}$$

For all $t = 1,2,3,\ldots,T$.

Where, St is reservoir storage at time $t(10^6)$ m^3; $S(t+1)$ is reservoir storage at next time period $t(10^6)$ m^3; In(t) is inflow to the reservoir at time period $t(10^6)$ m^3; $P(t)$

is release made to river bed turbine in period $t(10^6)$ m^3; IRR(t) is water release in period $t(10^6)$ m^3; Evp(t) is evaporation at time period $t(10^6)$ m^3; Ovf(t) is overflow at time period $t(10^6)$ m^3.

(b) Constraints of release from the reservoir

$$D_{\min}(t) \leq \mathrm{IR}_t \leq D_{\max}(t) \tag{22.15}$$

where $D_{\max}(t)$ is maximum irrigation demand (10^6) m^3; $D_{\min}(t)$ is minimum irrigation demand (10^6) m^3.

(c) Maximum power production limits

$$pP(t)H_t \leq E_{\max} \tag{22.16}$$

where p is power production coefficient; $P(t)$ is release made to river bed turbine in period $t(10^6)$ m^3; H_t is net heads available (m); $E_{\max}$ is maximum energy produced (KW).

(d) Constraints of industrial release from the reservoir

$$\mathrm{IND}_{\max} \leq \mathrm{INR} \leq \mathrm{IND}_{\min} \tag{22.17}$$

where $\mathrm{IND}_{\max}(t)$ is maximum industrial demand (10^6) m^3; $\mathrm{IND}_{\min}(t)$ is minimum industrial demand (10^6) m^3.

(e) Canal capacity limits

$$\mathrm{IRR}_t \leq \mathrm{Ca}_{\max} \tag{22.18}$$

where $\mathrm{Ca}_{\max}$ = Maximum canal capacity (10^6) m^3.

(f) Discharge from the reservoir through penstock

$$Q = A * V$$

$$= \frac{\prod}{4} d^2 \sqrt{2gH} \tag{22.19}$$

where Q is discharged from the reservoir for hydropower production (10^6) m^3; d is diameter of the penstocks (m); g is acceleration due to gravity; H is head for hydropower generation, and it is function of time (m).

22.5 Results and Discussion

A. PSO

At first, the numbers of iterations are arbitrarily chosen to start. By changing the iterations time to time, we are reaching to a point where optimality is maximum. For different iterations, water release for irrigation, power, and industry is found out (units-Million cubic meter), and corresponding efficiency (in %) is determined. For different constant values, the optimum values are checked by using MATLAB code (Fig. 22.2).

This graph shows the fitness value with respect to number of iterations. It is seen that after 167th iteration, the fitness value become constant. Particle swarm optimization has an advantage over DE since it is converged very fast as its convergence rate is high and time taken for convergence is less. After 167th iteration, its fitness value is converged to 2.280272.

B. TLBO

This figure represents of the fitness value for various iterations. With iteration, the fitness value is updated, and after some iteration, it attains to a constant value. For each iteration, the fitness value is recorded on the basis of that coding is progressed. From this figure, it is clearly seen that by using teaching–learning-based optimization, after 72nd iteration, it converges to one value. On the basis of fitness value, the effectiveness of algorithm is decided. Its fitness value varies from 256.2233 to 0.065159. After 72nd iteration, its fitness value is converged to 0.065159 (Fig. 22.3; Table 22.1).

For irrigation release for non-monsoon season is 134.28 Million m^3 and 137.26 Million m^3 by using PSO and TLBO, respectively. For hydropower generation for non-monsoon season are 189.515 and 207.315 MW by using PSO and TLBO.

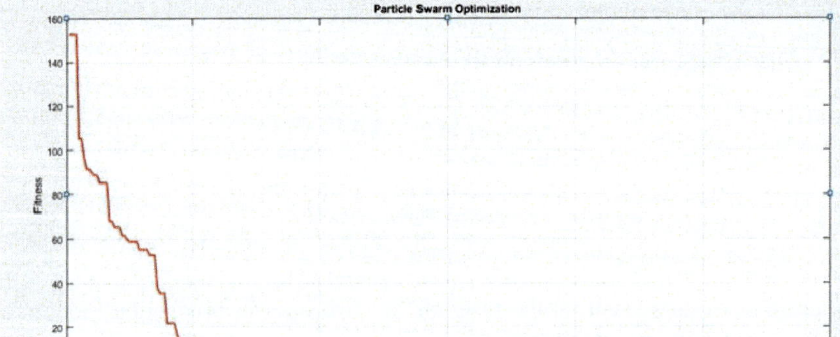

Fig. 22.2 Fitness value over iteration with particle swarm optimization

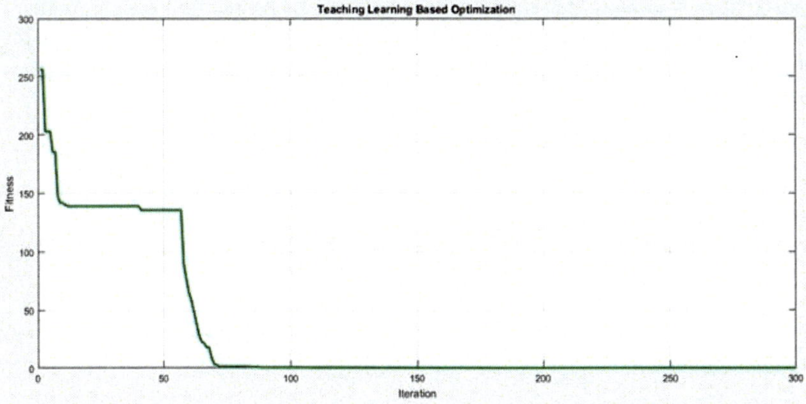

Fig. 22.3 Fitness value over iteration with teaching–learning-based optimization

Table 22.1 Comparison table showing the global optimum release by TLBO and PSO

Methods	TLBO(MCum)	PSO(MCum)
Water release for irrigation	137.26	134.28
Water release for power	697.00	692.6753
Water release for industries	82.156	78.481
Global optima	916.41	867.31

For industrial release for non-monsoon season are 78.481 and 82.156 Million m^3 by using PSO and TLBO (Tables 22.2, 22.3, and 22.4; Figs. 22.4, 22.5, 22.6, and 22.7).

This graph clearly represents that by using the teaching–learning-based optimization, we can conserve more end storage then compare to the actual end storage for a particular ten daily basis after the release of water from the reservoir for irrigation, power generation, and industrial purpose. Thus, the policy developed by teaching–learning-based optimization is far better than the policy of other techniques.

Table 22.2 Irrigation release for the historical years for the non-monsoon season from 1990 to 2016 for Hirakud reservoir system from TLBO, PSO and DE methods for 10-daily time periods (in Million m^3)

Statistical parameters	Models for reservoir operation	
	TLBO	PSO
Maximum value (Million m^3)	137.26	134.11
Minimum value (Million m^3)	25.78	20.69
Average value (Million m^3)	79.34	75.95

Table 22.3 Hydropower production for the historical years for the non-monsoon season from 1990 to 2016 for Hirakud reservoir system from TLBO, PSO, and DE methods for 10-daily time periods (in MW)

Statistical parameters	Models for reservoir operation	
	TLBO	PSO
Maximum value (MW)	207.315	189.515
Minimum value (MW)	1.369	1.178
Average value (MW)	32.387	29.441

Table 22.4 Industrial release for the historical years for the non-monsoon season from 1990 to 2016 for Hirakud reservoir system from TLBO, PSO and DE methods for 10-daily time periods (in Million m^3)

Statistical parameters	Models for reservoir operation	
	TLBO	PSO
Maximum value (Million m^3)	82.156	78.15
Minimum value (Million m^3)	11.25	10.25
Average value (Million m^3)	80.09	79.73

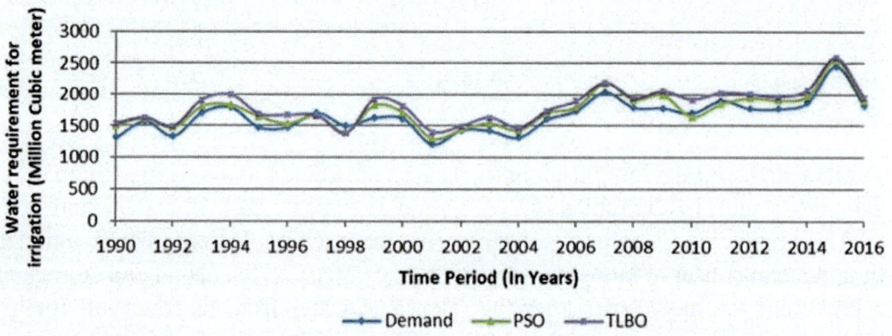

Fig. 22.4 Comparison of irrigation release for 27 years by using optimization techniques

22.6 Conclusions

This study developed optimal operation policies by using soft computing methods for Hirakud reservoir system. On applying evolutionary algorithms such as TLBO, PSO, and DE for solving reservoir operation problems, it is found that the TLBO method is resulting in better performance for reservoir optimization by providing good quality optimal solutions in fewer numbers of iterations as compared to other two methods. The following conclusions are drawn from various experiments done for 27 years that for irrigation release for non-monsoon season are 134.28 Million m^3 and 137.26 Million m^3 by using PSO and TLBO, respectively, for hydropower generation for non-monsoon season are 189.515 MW and 207.315 MW by PSO and TLBO,

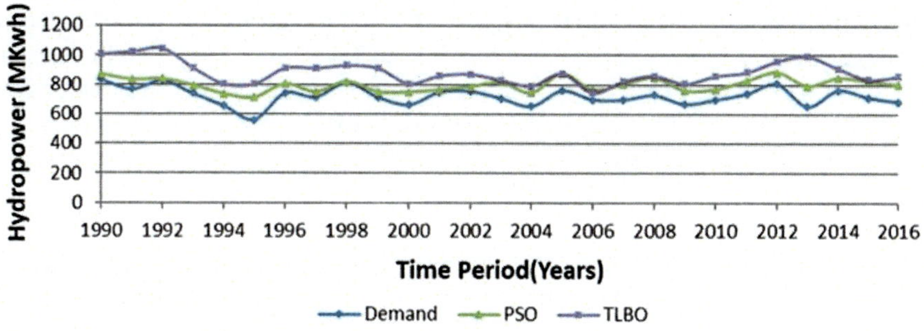

Fig. 22.5 Comparison of hydropower generation for 27 years by using optimization techniques

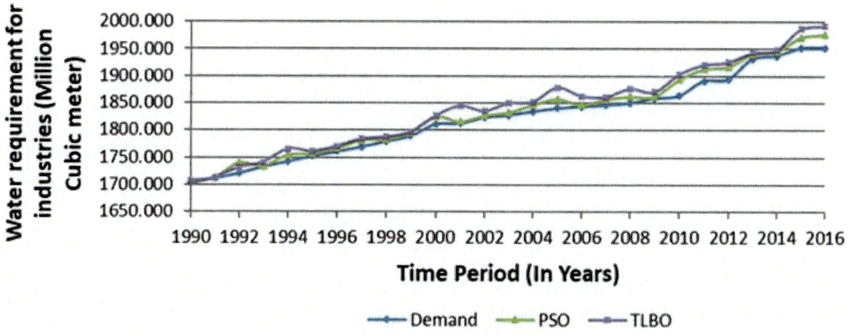

Fig. 22.6 Comparison of industrial release for 27 years by using optimization techniques

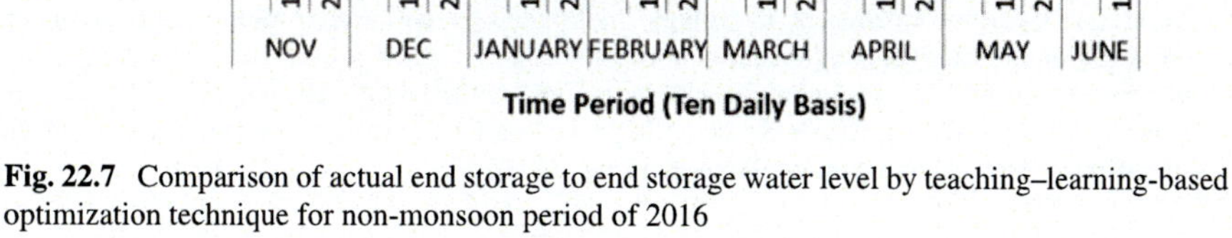

Fig. 22.7 Comparison of actual end storage to end storage water level by teaching–learning-based optimization technique for non-monsoon period of 2016

respectively, and for industrial release for non-monsoon season are 78.481 Million m^3 and 82.156 Million m^3 by using PSO and TLBO, respectively. For the single year 2016, the actual release for the irrigation is 130.683 Million m^3, but for PSO and TLBO is 133.9175 Million m^3 and 134.4254 Million m^3, respectively, whereas the actual hydropower generation are 176.3101 MW, but for PSO and TLBO is

181.3101 MW and 185.949 MW, respectively. And the actual releases of water for industries are 72.683 Million m^3, but for PSO and TLBO are 75.9175 Million m^3 and 80.4254 Million m^3, respectively. Therefore, on comparing the actual release and the policy developed by PSO and TLBO methods for reservoir operation problems, it is obtained that the release of water for irrigation, power generation, and industrial purpose is more by using TLBO than that of the release by other techniques.

Acknowledgements I would like to express my heartfelt gratitude to Prof. Prakash Chandra Swain, my supervisor for rendering their valuable guidance and suggestions, encouragement, comments and constructive criticisms given then and there which has made this research work a successful one. Without his kind help, I would not even have a chance to work on this area. Furthermore I am greatly indebted to EE, Hydrology Division, Hirakud Dam Circle for sharing the real-life data.

References

Eberhart RC, Kennedy J (1995) A new optimizer using particle swarm theory. In: Proceedings of 6th symposium on micro machine and human science. IEEE Service Center, Piscataway, NJ, pp 39–43

Kumar DN, Raju KS, Ashok B (2006) Optimal reservoir operation for irrigation of multiple crops using genetic algorithms. J Irrig Drainage Eng 132:123–129

Rao RV, Savsani VJ, Vakharia DP (2011) Teaching–learning-based optimization: a novel method for constrained mechanical design optimization problems. Comput Aided Des 43:303–315

Rao KP, Savsani VJ (2012) Mechanical design optimization using advanced optimization techniques. Springer, London

Rao RV, Savsani VJ, Balic J (2012a) Teaching–learning-based optimization algorithm for unconstrained and constrained real parameter optimization problems. Eng Optim 44(12):1447–1462

Rao RV, Savsani VJ, Vakharia DP (2012b) Teaching–learning-based optimization: a novel optimization method for continuous non-linear large scale problems. Inf Sci 183(1):1–15

Rao KP, Teja RB (2015) Teaching learning based optimization for short term hydro-thermal scheduling. Conference on power, control, communication and computational technologies for sustainable growth (PCCCTSG) 978-1-4673-6890-2/15

Reddy MJ, Kumar DN (2009) Performance evaluation of elitist-mutated multi-objective particle swarm optimization for integrated water resources management. J Hydroinformatics 79–88

Swain PC, Nanduri UV (2009) Neuro-fuzzy inference system for operation of a multi-purpose reservoir. Hydroinformatics in hydrology, hydrogeology and water resources. International Association of Hydrological Sciences Publication 331

The project report on Hirakud dam (2016)—by master control station, Burla

Zhang J, Wu Z, Cheng C, Zhang S (2011) Improved particle swarm optimization algorithm for multi-reservoir system operation. Water Sci Eng 4(1):61–73

Chapter 23
Agricultural Water Management and Groundwater Recharging Using Vadose Zone Modelling

Anooja Thomas, Vivekanand Singh, and Brijesh Kumar Yadav

Abstract Temporal and spatial soil moisture status in variable-saturated zone plays a vital role in agricultural water management, groundwater recharge and solute transport of that region, and it is fundamentally regulated by plant growth and its water extraction rate along with soil properties and weather conditions. The soil moisture distribution pattern at soil element scale is regulated by Richard's equation integrated with a sink term representing water uptake rate by plants. For obtaining soil moisture dynamic across the root zone of a site-specific area, Richards equation-based model coupled with sink term is numerically solved using finite-difference method. The developed model is tested for simplified case(s) and applied over a wheat-cropped site in Patna region of Bihar, India, with pedotransfer functions obtained from soil sample analysis. Stimulation results for a period of 120 days show that out of 24 cm irrigation applied and 10.61 cm rainfall occurred, 47.2% has been lost as evapotranspiration, 28.9% of the water has been stored in the soil column as soil moisture, and 23.9% of the water has recharged to groundwater table. Based on these results, an irrigation schedule is recommended for an optimal utilization of both rainwater and irrigation water for a better management of available water resources.

Keywords Groundwater recharge · Root water uptake · Soil moisture modelling · Numerical modelling

A. Thomas (✉) · B. K. Yadav
Department of Hydrology, Indian Institute of Technology Roorkee, Roorkee, Uttarakhand 247667, India

B. K. Yadav
e-mail: brijkfhy@iitr.ac.in

V. Singh
Department of Civil Engineering, National Institute of Technology Patna, Patna, Bihar 800005, India
e-mail: vsingh@nitp.ac.in

R. Jha et al. (eds.), *Water Resources Management and Reservoir Operation*,
Water Science and Technology Library 107,
https://doi.org/10.1007/978-3-030-79400-2_23

23.1 Introduction

The compilation of World Bank experiences and strategies stated that water management of rain-fed agriculture is one of the major challenges of irrigated agriculture. The cultivation practices in India cannot vary dependent on unpredictable monsoon. Excessive irrigation results in water logging and soil salinity in addition to pesticides and nutrient leaching groundwater contamination (Yadav and Junaid 2014) and water stress affects yield during its sensitive growing stage. Moreover, the study area makes use of arsenic-contaminated groundwater for the irrigation purpose. This may lead to major health risk in near future (Singh and Ghosh 2011). Soil moisture status serves as a tool for the management of the irrigation and thereby for a cost effective, better crop yield and for considering other agriculture-related environmental issues.

The soil moisture flow through the partially saturated soil is a complex system governed by highly nonlinear processes described by means of physical–mathematical models. Modelling progress is achieved by one dimensional Richard's equation for describing infiltration through vadose zone which is coupled with sink term for the account of evapotranspiration (ET) and plant growth stage (Yadav et al. 2009). Root water uptake is a spatial and temporal process influenced by plant type, soil characteristics and the prevailing climatic conditions, plant growth stage, rooting depth and density distribution (Yadav and Mathur 2008) and availability of soil water. Numerous methods have proposed with various degree of accuracy for the estimation of root water uptake. There are major three approaches for computing root water uptake, microscopic, mesoscopic and macroscopic approaches. Microscopic approach simulates individual root considering individual root as a uniform radius, long cylinder with water-absorbing properties (Feddes and Raats 2004; Šimůnek and Hopmans 2009). The mesoscopic approach simulates the root water uptake through diffusion and is the function of matrix flux potential difference of the soil and the root (van Lier et al. 2006).

Numerous water extraction models are proposed over macroscopic empirical model (Feddes et al. 1978; Ojha and Rai 1996; Yadav et al. 2009). Water uptake patterns proposed can categorize as constant (Feddes et al. 1978), linear, (Prasad 1988), nonlinear (Shankar et al. 2012) or exponential (Li et al. 1999; Kang et al. 2001). Molz and Remson (1970) proposed a hypothetical linear pattern from four layer of equal depth with water uptake of 40, 30, 20 and 10% top from each layer.

Groundwater recharge through the unsaturated zone of a cropped site has simulated considering that the excess water after the soil moisture holding capacity of soil, which drains below the root zone and reaches the groundwater table (Rushton 1988; Wohling et al. 2012). Numerous studies are focused on the potential grounder estimation through unsaturated zone the focus of study (Jimenez-Martinez et al. 2009; Min et al. 2015) as it required less parameterization effort calculation efficiency, cost effective and its predicting capability compared to saturated zone and surface water-based models.

This study focused on developing a simplified and least parameterized soil moisture model for the recharge estimation along with agriculture water management and its field application over a wheat-cropped site at Patna, Bihar.

23.2 Methodology

Special and temporal movement of soil moisture through the unsaturated zone is governed by the Richard's equation coupled with the sink term for the account of the out flux of soil moisture from the system through plants. The mixed form of Richard's 1D equation is written as,

$$\frac{\partial \theta}{\partial t} = \frac{\partial}{\partial z}\left[K(\psi)\left(\frac{\partial \psi}{\partial z} - 1\right)\right] + S(z, t) \tag{23.1}$$

where θ = volumetric soil water content (volume of water per unit volume of soil); ψ = soil matrix potential; $S(z, t)$ = sink term representing water extraction by plant roots from a unit volume of soil per unit time; z = vertical coordinate (taken positive upwards); K = unsaturated hydraulic conductivity of the soil; and t = time. For solving mixed form of Richard's equation, it requires functional relation between soil moisture content θ and hydraulic conductivity K in term of soil moisture potential, ψ. Soil moisture characteristics relationship proposed by Haverkamp et al. (1977) is used.

23.2.1 Initial and Boundary Conditions

For the initialization of the simulation, the initial condition of the soil moisture condition or water potential corresponding to field condition at the starting time of simulation is specified at each node. The actual field condition of soil moisture status at individual is difficult to measure. For this reason, an arbitrary value of soil moisture content is selected. Corresponding pressure head and unsaturated hydraulic conductivity (computed using the characteristic relationships) are also specified at all the nodes as the initial conditions,

$$\text{i.e.}\quad \theta(z, t) = \theta_{\text{ini}} \quad \text{for} \quad z \geq 0, t = 0 \tag{23.2}$$

The depth of irrigation and rainfall and evaporation from the surface is defined as the upper boundary condition. For the subsurface flow resulting from irrigation and rainfall infiltration, the upper BC will change with time. During the initial stage of irrigation/rainfall, a specified depth of ponding ψ $(x) = h(x)$ (= depth of water) is considered. Let the imposed pressure head at the surface of ground for the flow domain be ψ_b. The above pressure head ψ_b is used along with the pressure head of first node ψ_1 and pressure head at the second node ψ_2 in the direction of z-axis for the determination of the ground surface flux as given below,

$$V_z|_{z=0} = -K(\psi_b)\left\{\frac{(-8\psi_b + 9\psi_1 - \psi_2)}{3\Delta z}\right\} \tag{23.3}$$

As time progresses, the water gets infiltrated into soil, which starts decreasing of flow depth at the surface. When all water gets infiltrated into soil, then the upper boundary is changed to the flux-type boundary. The flux boundary is taken as the irrigation and rainfall infiltration and actual rate of evaporation at the first node near the ground surface with a condition that it cannot falls below $\psi_{\min}$, where

$$\psi_{\lim} = \frac{RT(t)}{Mg}\ln[f(t)] \tag{23.4}$$

where $\psi_{\min} = \psi_b$ is the pressure head at the surface (bar); $R =$ the universal gas constant (8.31×10^7 eg/mole/K); $T(t)$ is the absolute temperature (0 K); $M =$ molecular weight of water (18 gm/mole); $g =$ acceleration due to gravity.

23.2.2 Root Growth Function and Potential Evapotranspiration

Atmospheric boundary condition requires specifying rate of potential evapotranspiration, irrigation and precipitation. Potential evapotranspiration is calculated using pan evaporation data. For more accurate rather than considering pan evaporation method, evaporation is calculated using Christiansen (1968). Pan evaporation data is related to the potential evapotranspiration (E_{vpt}) using equation $E_{\text{vpt}} = KE_p$ as suggested by Allen et al. (1998), where E_p is the pan evaporation, and K is the consumptive use coefficient, which depends upon the type of crops and growth stages and characterizes evaporation and plant water uptake relative to pan evaporation data. The distribution of the potential evapotranspiration alone the root depth is given as,

$$E_{\text{pdis}}(z) = C_{\text{rt}}E_{\text{vpt}}(0,t) \quad \text{For} \quad 0 \le z \le d \tag{23.5}$$

where $E_{\text{pdis}}(z) =$ distribution of potential evapotranspiration based on root density; $C_{\text{rt}} =$ coefficient of the root density at depth z; and $d =$ depth of the root zone. The root growth function or coefficient of root density is considered as the function defined by Molz and Remson (1970).

The actual evapotranspiration is calculated based on the available soil moisture at that node and is given as (Zaradny 1993):

$$E_a(z,t) = \begin{cases} E_{\text{pdis}}\left(\frac{\theta-\theta_w}{\theta_f-\theta_w}\right) & \text{For } -15800\text{cm} \le \psi \le -400\text{cm} \\ E_{\text{pdis}} & \text{For } -400\text{cm} \le \psi \le -50\text{cm} \\ E_{\text{pdis}}\left(\frac{\theta_s-\theta}{\theta_s-\theta_{\text{an}}}\right) & \text{For } -50\text{cm} \le \psi \le 0\text{cm} \end{cases} \tag{23.6}$$

where $E_a(z, t)$ is the actual evapotranspiration from the root of the crop (cm) at particular node; θ_w is the soil moisture at wilting point; θ_f is the soil moisture content at field capacity; θ_{an} is the soil moisture content at anaerobiosis point; and θ_s is the saturated moisture content. The variation of actual evapotranspiration (AET) along the root zone depth is given by (Michael 1978)

$$S_t(0,t) = \begin{cases} 0.4 \times E_a(0,t) \text{ For } 0 \le z \le \frac{d}{4} \\ 0.3 \times E_a(0,t) \text{ For } \frac{d}{4} \le z \le \frac{d}{2} \\ 0.2 \times E_a(0,t) \text{ For } \frac{d}{2} \le z \le \frac{3\times d}{4} \\ 0.1 \times E_a(0,t) \text{ For } \frac{3\times d}{4} \le z \le d \end{cases} \tag{23.7}$$

23.2.3 Laboratory Experiment and Parameter Estimation

The model simulation requires to specify the parameters θ_s, θ_r and K_s. Five soil samples were collected from the field and tested in the equipment, pressure plate apparatus for the soil moisture retention curve. Soil moistures corresponding to applied pressure are recorded for the laboratory observed soil moisture retention curve. Pedotransfer function model, RETC Code (van Genuchten et al. 1991), analyses the hydraulic conductivity functions and soil water retention of unsaturated soil from subset of the measured pressure head and water content data. The observed pressure head vs. soil moisture data provided as the input to the RETC Code and obtained the parameters soil moisture retention curve θ_s, θ_r and K_s. The simulated soil moisture retention curve fitted in MATLAB and obtained the Haverkamp constants, m and n. The soil samples are subjected to particle size analysis for the soil type. The estimated parameters are given in Table 23.1.

Table 23.1 Hydraulic properties of soil sample

Hydraulic properties	θ_s	θ_r	A	m	B	n
Loam soil	0.3460	0.038	121.1	2.446	29.45	0.8011

23.2.4 Simulation Parameters

Daily climatic data including pan evaporation, rainfall, mean temperature, wind velocity, relative humidity and mean sunshine hour have been collected from ICAR Patna near wheat-cropped site from 21th November to 20th March for the duration of 120 days. Accurate pan evaporation value calculated using Christiansen (1968) and rainfall event is shown in Fig. 23.1. The simulation domain parameters and the crop parameters are explained in the Tables 23.2 and 23.3 respectively.

23.2.5 Numerical Method

Soil moisture dynamics of vadose zone is simulated using 1D Richard's equation, and water extraction pattern is calculated according to the macroscopic approach

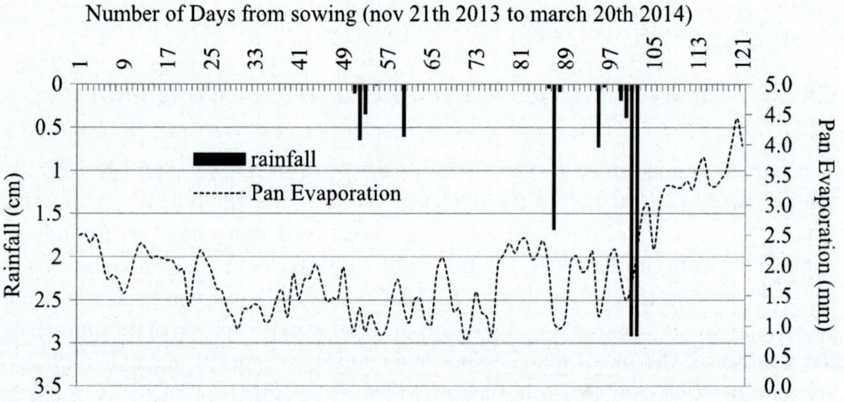

Fig. 23.1 Daily pan evaporation, precipitation and irrigation for the entire crop period

Table 23.2 Domain parameters used in simulation of soil moisture regime in wheat-cropped site

Domain parameters	Details
Node size	2 cm
Depth of simulation	200 cm
Initial condition	−40 cm (suction head at field capacity)
Upper boundary condition	Both flux and head boundary condition
Lower boundary condition	Groundwater table
Field capacity	−40 cm
Permanent wilting point	−15,800 cm

Table 23.3 Crop parameters used in simulation for wheat crop

Crop parameters	Details
Crop type	Wheat
Rooting depth	60 cm
Crop duration	120 days (Initial growth stage 0–20 days Development stage—20–40 days Mid-season 40–90 days Late season 90–120 days)
Irrigation details	Four irrigations with 6 cm depth each (On 21th, 60th, 90th and 110th days)

suggested by Molz and Remson (1970) due to its simplicity and requirement of less number of model parameters and solved numerically along with an appropriate initial and boundary conditions. The solution of the partial difference equation is obtained using the strongly implicit finite-difference scheme (Hong et al. 1994; Singh and Bhallamudi 1998). The equation is specified at each node, and numerical technique is used to solve the nonlinear equation in each time steps. At each iteration process, a set of algebraic equations in tridiagonal matrix form are formulated. The iterative process will continue to obtain soil moisture content until all nodes reach a reasonable degree of convergence. If the soil moisture of a particular node is more than the field capacity, it will transfer to the next node below it and finally added up to get groundwater recharge. A programming code is developed in FORTRAN for solving the governing equations.

23.3 Results and Discussion

The model is simulated for a simplified case without root uptake and compared well with the available literature data. The model is then applied to a wheat field for predicting soil moisture regime sand subsequent groundwater recharge.

23.3.1 Soil Moisture Flow Without Root Water Uptake

The model is first validated over the soil sample simulated by Haverkamp et al. (1977) without sink term. The input parameters for the characteristics relationship were saturated hydraulic conductivity, $K_s = 34$ cm/hr; saturated moisture content, $\theta_s = 0.287$; residual moisture content, $\theta_r = 0.075$; the constants $A = 1.175 \times 10^6$; $m = 4.74$; $B = 1.611 \times 10^6$; and $n = 3.96$. The depth of the soil column was 80 cm. The initial condition was specified as the volumetric moisture content, $\theta_{\text{ini}} = 0.1$. The boundary condition at the top was considered as the constant flux = 13.69 cm/hr, and the bottom boundary condition was specified as the constant moisture contents 0.1

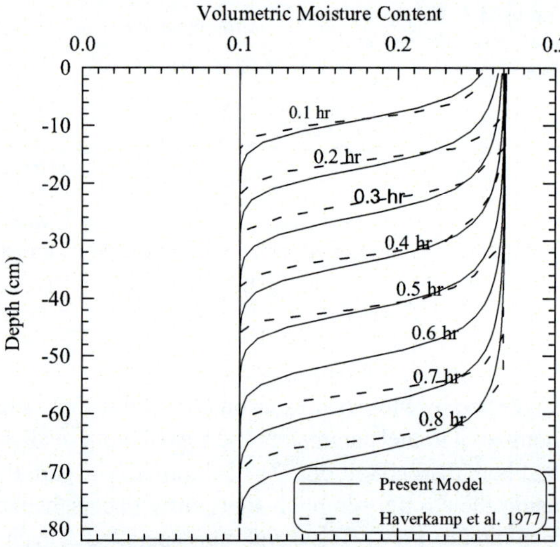

Fig. 23.2 Verification of the model with experimental data: soil moisture profiles dashed lines for experimental and solid lines for present model

(Haverkamp et al. 1977). Figure 23.2 shows the movement of moisture content with different times computed with present model and Haverkamp et al. (1977). Results of both the models are very close.

23.3.2 Soil Moisture Flow Through the Vadose Zone

The soil moisture movement through the soil after the application of the irrigation on 21th day of crop period is presented in Fig. 23.3. Soil moisture moves slowly down through the soil to reach the entire soil column into its field capacity. Soil moistude content above the field capacity add as the recharge to the groundwater. Eventually, the soil moisture decreased due to the out flux in the form of root water uptake.

23.3.3 Prediction of Irrigation Scheduling

During the early days of irrigation, the soil moisture is abundant, and the water uptake is at its optimal condition. The loss of soil moisture from top layer by evaporation and water uptake by plant roots decreases the soil moisture status eventually depending on the root density distribution. At this stage, root water uptake will be in its potential rate. Once the soil moisture content falls down to wilting point, further extraction from that node is zero, and evapotranspiration reduces than its potential value. Prolonged water-stressed condition results in the wilting of the plants, and gradually, permanent wilting occurs. This will affect the yield of the crop.

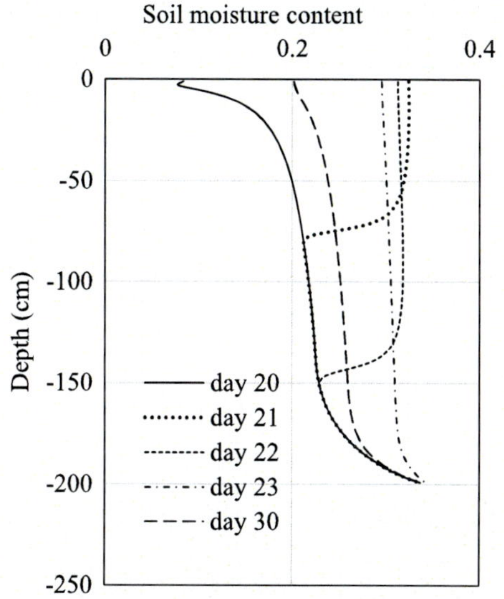

Fig. 23.3 Movement of soil moisture flux through the soil column under saturated bottom boundary condition

Due to continuous loss of soil moisture from the top layer, the availability decreased, and the actual evapotranspiration becomes less than its potential value. From Fig. 23.4, it can be observed that actual evapotranspiration is less than its potential during the 38th–60th day and from 76 to 84th day. Irrigation is recommended during this period.

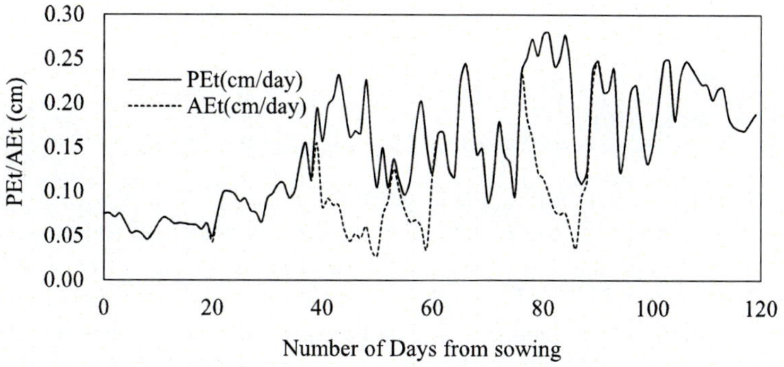

Fig. 23.4 Potential evapotranspiration (PEt) and actual evapotranspiration (AEt) of the wheat crop

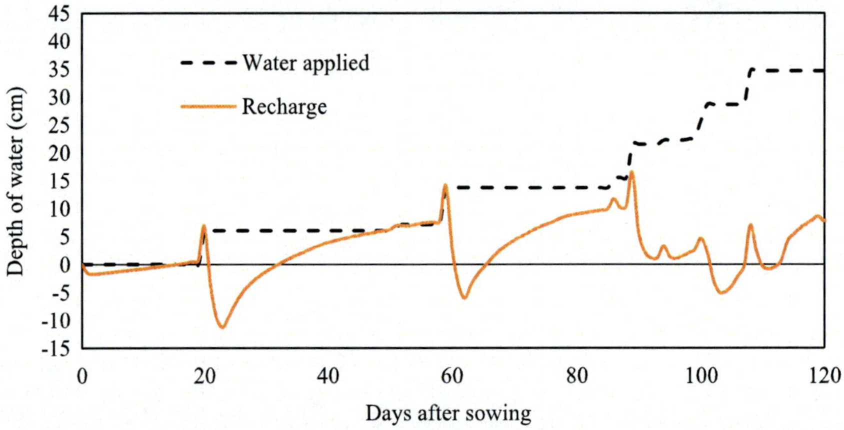

Fig. 23.5 Recharge of groundwater over the crop peroid

23.3.4 Groundwater Recharged for the Wheat Crop Region

The developed model is used for recharge estimation in an agricultural field with loam soil. Over the entire crop period of 120 days (21st November 2013–20th March 2014), total 10.614 cm of rainfall has occurred, and 24 cm of irrigation water has been applied. Thus, total water application is 34.614 cm. In early irrigation, all the water was built as the soil moisture in the soil column, and some water has added from the groundwater to soil. It is shown as negative sign in Fig. 23.5. During the entire crop period, total 8.26 cm of water has been recharged, 16.25 cm has been lost as evapotranspiration, and 10.01 cm of water has been stored in the soil column as soil moisture.

23.4 Conclusions

A 1D Richard's equation-based soil moisture model has been developed through this study for the prediction of temporal and spatial movement of soil moisture after irrigation or rainfall and the contribution to underlying groundwater. Sink term is coupled with mixed form of Richard's equation for the account of root water uptake by plants variation. The nonlinear Richard's equation is solved numerically using highly implicit finite-difference method, and iteration technique is used for predicting unknown soil moisture and suction head. The developed model is validated with results of Haverkamp et al. (1977). Results of present model are very close to the Haverkamp's results.

The model has been used to simulate a wheat-cropped site with loam soil in Patna district Bihar. The daily soil moisture regime, suction head and actual evapotranspiration are simulated for the entire crop period. During the crop period from 21th November to March 20th, total 10.614 cm of rainfall has occurred, and 24 cm of irrigation water has been applied. Out of 34.6 cm of total water applied, 23.9% of water has recharged, 47.2% has been lost as evapotranspiration, and 28.9% of water has been stored in the soil column as soil moisture. The model is strictly following water balance, and hence, it is justifiable. From the simulation, it is noted that the actual evapotranspiration is less than its potential rate during 38th–60th days and from 76 to 84th days. Thus, additional irrigation is recommended during this periods.

The model developed can be applied at any land area if the weather data, soil type and cultivation details are available. Thus, the advantage of this model is that it can be used for recharge estimation of a cultivated area, and it can be obtained without visiting a field or detailed data. It can also give the information of irrigation requirements of the crop. The irrigation schedule can be planned with the help of soil moisture profile, and the depth of water required for the irrigation also can be determined.

References

Allen RG, Pereira LS, Raes D, Smith M (1998) Crop evapotranspiration. Guidelines for computing crop water requirements. Irrigation and drainage paper no. 56, FAO. Rome, Italy

Christiansen JE (1968) Pan evaporation and evapotranspiration from climatic data. J Irrig Drainage Eng, ASCE 94(2):243–265

Feddes RA, Raats PAC (2004) Parameterizing the soil–water–plant root system. In: Feddes RA, de Rooij GH, van Dam JC (eds) Proceedings of the unsaturated zone modelling: progress, challenges and applications. Wageningen UR Frontis Series, vol 6. Kluwer Academic Publishers, Dordrecht, The Netherlands, x–xi (Chapter 4), pp 95–141

Feddes RA, Kotwalik PJ, Zaradny H (1978) Simulation of field water use and crop yield. Centre for Agricultural Publishing and Documentation, Wageningen, Netherlands

Haverkamp R, Vauclin M, Touma J, Wieranga PJ, Vachaud G (1977) A comparison of numerical simulation models for one dimensional infiltration. Soil Sci Soc Am J 41:285–294

Jimenez-Martinez J, Skaggs TH, van Genuchten MT, Candela L (2009) A root zone modelling approach to estimating groundwater recharge from irrigated areas. J Hydrol 367(1–2):138–149

Kang S, Zhang F, Zhang J (2001) A simulation model of water dynamics in winter wheat field and its application in a semiarid region. Agric Water Manage 49:115–129

Li KY, Boisvert JB, De Jong R (1999) An exponential root water uptake model. Can J Soil Sci 79:333–343

Michael AM (1978) Irrigation: theory and practice. Vikas Publishing House PVT Ltd

Min L, Shen Y, Pei H (2015) Estimating groundwater recharge using deep vadose zone data under typical irrigated cropland in the piedmont region of the North China plain. J Hydrol Eng 527:305–315

Molz FJ, Remson I (1970) Extraction term models of soil moisture use by transpiring plants. Water Resour 6(5):1346–1356

Ojha CSP, Rai AK (1996) Nonlinear root water uptake model. J Irrig Drainage Eng 122:198–202

Prasad R (1988) A linear root water uptake model. J Hydrol 99:297–306

Rushton K (1988) Numerical and conceptual models for recharge estimation in arid and semi-arid zones. Estimation of natural groundwater recharge. Springer, pp 223–238

Shankar V, Hari Prasad K, Ojha C, Govindaraju R (2012) Model for nonlinear root water uptake parameter. J Irrig Drainage Eng 10:1061

Singh V, Bhallamudi SM (1998) Conjunctive surface subsurface modelling of overland flow. Adv Water Resour 21(7):567–579

Singh SK, Ghosh AK (2011) Entry of arsenic into food material—a case study. World Appl Sci J 13(2):385–390

Šimůnek J, Hopmans JW (2009) Modeling compensated root water and nutrient uptake. Ecol Model 220(4):505–521

van Genuchten MTh, Leij FJ, Yates SR (1991) The RETC code for quantifying the hydraulic functions of unsaturated soils. USEPA, Washington, DC

van Lier QDJ, Metselaar K, van Dam JC (2006) Root water extraction and limiting soil hydraulic conditions estimated by numerical simulation. Vadose Zone J 5:1264–1277

Wohling DL, Leaney FW, Crosbie RS (2012) Deep drainage estimates using multiple linear regression with percent clay content and rainfall. Hydrol Earth Syst Sci 16(2):563–572

Yadav B, Junaid S (2014) Groundwater vulnerability assessment to contamination using soil moisture flow and solute transport modeling. J Irrig Drainage Eng 10:1061

Yadav BK, Mathur S (2008) Modeling soil water uptake by plants using nonlinear dynamic root density distribution function. J Irrig Drainage Eng 134(4):430–436

Yadav BK, Mathur S, Siebel MA (2009) Soil moisture flow modeling with water uptake by plants (wheat) under varying soil and moisture conditions. J Irrig Drainage Eng 135(3):375–381

Zaradny H (1993) Groundwater flow in saturated and unsaturated soil. Zeidler RB (ed) Balkema, Rotterdam, The Netherlands